JCT 2005:
Clause by Clause

Phil Griffiths

ELSEVIER

Amster   Oxford
Paris ·   · Tokyo

BH

Butterworth-Heinemann is an imprint of Elsevier
The Boulevard, Langford Lane, Kidlington, Oxford OX5 1GB, UK
30 Corporate Drive, Suite 400, Burlington, MA 01803, USA

First edition 2010

British Library Cataloguing in Publication Data
A Catalogue record for this book is available from the British Library

Library of Congress Cataloging-in-Publication Data
A Catalog record for this book is available from the Library of Congress

ISBN–13: 978-1-8561-7518-0

For information on all Butterworth-Heinemann publications
visit our website at www.books.elsevier.com

Printed and bound in Great Britain

10 11 12 13 14   10 9 8 7 6 5 4 3 2 1

Contents

Preface

During the early years of the 21st century, the Joint Contracts Tribunal (JCT) decided to update and modernise their collection of construction contracts. Consequently a number of revamped contracts were published during 2005, with more following over the next few years. As part of their review, the JCT took the opportunity to employ a new uniform style and structure in the majority of their contracts. The contract conditions were largely unchanged from the previously published contracts, although some significant changes were made in response to current commercial and contractual practices.

Part 1 of the book undertakes a review of these updated JCT contracts. The purpose of the review is to identify the type of construction work for which each contract is best suited, to examine the format and content of the contract and, finally, to compare and contrast the contract conditions with those of the JCT Standard Building Contract with Quantities (SBC/Q). This is not intended to be an in-depth analysis and should be viewed as an introduction to the suite of contracts currently published by the JCT.

By contrast, Part 2 takes an in-depth look at one of the JCT's standard forms of contract, i.e. the SBC/Q. This part of the book comprises a clause-by-clause analysis of the contract and provides an explanation as to the rationale and operation of each clause. The purpose of this section is to act as an aid in the interpretation and understanding of the sometimes complex language found in the SBC/Q. To assist in this process, the layout of this section deliberately mirrors the JCT style and format so that the text may be easily compared with the SBC/Q and vice versa. Although the use of the SBC/Q has diminished in recent years, it still provides a solid basis for good management and contract administration. Many of its clauses and procedures are to be found in the other JCT contracts and, as a consequence, a good understanding of the SBC/Q should lead to an improved understanding of contract administration in general.

The lesson to be learnt from Part 3 is that, no matter how much care is taken by legal draftsmen in preparing contract conditions, situations will often arise where two parties will read the same clause and arrive at conflicting interpretations. The courts may be used to resolve an argument of interpretation, but even they struggle at times to unravel the complex contract wording before they can arrive at the 'correct' interpretation. However, the courts' decisions can often be of assistance in determining how a clause should be read and how it should operate. As a consequence, Part 3 of the book reviews a number of legal cases that have helped to shape the current JCT contracts and assisted in the interpretation of some of the more complex areas of the contract conditions.

Phil Griffiths

March 2009

PART 1

Contract Documentation

W here a building project is to be undertaken, the client will have to engage a considerable number of individuals and organisations to provide him with advice, services, materials and goods. To ensure that the work progresses smoothly, it is essential that all the parties involved are fully aware of their role within the project and have clear guidelines as to their rights and obligations throughout the project. This may be achieved through the use of a series of suitable contractual agreements.

A contract may come into existence through a verbal agreement, an exchange of letters, standard conditions of trading or a standard form of contract. All these agreements are legally enforceable, but arguably some are more so than others. For example, with a verbal agreement, who is to know what was really agreed between the parties, and how will the passage of time affect people's memories and recollections?

In most instances, it is not a legal requirement for any contract to be evidenced in writing unless the parties have agreed to enter into a contract as a deed. But, for obvious commercial reasons, the vast majority of contracts will be entered into by using a written form of contract. The advantage of having a written form of contract is that there will be express terms written into the document to lay down clearly the rights and obligations of the parties involved; therefore, if a dispute arises during the progress of a project, it should be possible to determine who is at fault and what procedures should be followed. If necessary, the contract conditions would also provide the key details in the event of a legal action for breach of contract.

The vast majority of major building works are carried out in accordance with the terms and conditions of a written form of contract, and frequent use is made of published standard forms of contract. Over the years, there has been a steady growth in the development and publication of standard forms of contract, a fact that reflects, to some extent, the increasing diversity of procurement methods, which clients are now prepared to utilise (e.g. design and build, management fee, fast-track, partnering, guaranteed maximum price, target cost). The choice of contracts can at times be bewildering, and this can be compounded by a failure to understand and interpret complex contract conditions and procedures. As a consequence, there are a number of parties who would like to see contract documentation simplified.

Currently, there are two organisations producing the majority of standard forms of contract used on UK building projects. These organisations are the Joint Contracts Tribunal (JCT), which produces the aptly named JCT forms of contract, and the Institute of Civil Engineers (ICE), which is responsible for the New Engineering Contract (NEC) suite of contracts. There are other organisations that also produce standard forms of contract for building works but they are not widely used. The JCT and ICE have adopted contrasting methods of producing standard forms of contract to meet the varying demands of the construction industry. The JCT has taken the approach of producing a specific standard form of contract for each type of procurement route that may be selected, whereas the ICE, through the publication of the NEC documentation, has supported a move towards a more simplified, standardised and user-friendly form of contract documentation.

The NEC documentation comprises a set of core clauses, which will apply to any chosen procurement route. The client then selects from one of the six main procurement options (i.e. options A to F) and finally he has a choice of a further 20 optional clauses, any of which may be selected for incorporation into the contract. By opting for the NEC approach, it is possible to build up a collection of contract documentation suitable for any procurement approach the client may wish to adopt, and through the use of core clauses (which will apply to all NEC contracts), it becomes easier for parties within the construction industry to become familiar with the NEC style of contract administration.

It is interesting to note that this style of contract administration was being championed as far back as the mid-twentieth century in the Banwell Report[1]:

> *A multiplicity of conditions of contract for use in circumstances which vary only to a limited degree is not in our view conducive to a ready understanding between the parties to contract of their respective rights and obligations, and causes much work in checking variants and making adjustments. Many have expressed views in favour of a single form of contract for all building and civil engineering work…we believe that a common form is both possible and practicable.*

Furthermore, in 1994, Sir Michael Latham gave strong support to the NEC, although he did suggest that a number of alterations should be made for it to satisfy his guidelines on what comprises an effective form of contract. One minor suggestion was that its name should be changed to the 'New Construction Contract' to show clearly that it was adaptable for use on both engineering and building projects. Consequently, in 1995, the main contracts for use with the NEC were renamed as the 'Engineering and Construction Contracts'.

Despite the support for a common form of contract, there is a counterargument to this approach. It may be argued that a set of common documentation cannot fully take into account all the nuances and requirements of each different procurement route. As a result, the common documentation may provide a satisfactory contract package but it may not always provide the level of detail that some clients require. This may be why the JCT has adopted a significantly different approach to its contract documentation compared with the NEC.

The policy of the JCT is to produce a discrete standard form of contract for each type of project or procurement route commonly used in the UK. As a result, the JCT publishes a large selection of standard forms of contract and documentation as listed below:

Main Contracts

Standard Building Contract with Quantities (SBC/Q)
Standard Building Contract without Quantities (SBC/XQ)
Standard Building Contract with Approximate Quantities (SBC/AQ)
Design and Build Contract (DB)
Major Project Construction Contract (MP)
Management Building Contract (MC)
Constructing Excellence Contract (CE)
Intermediate Building Contract (IC)
Intermediate Building Contract with contractor's design (ICD)
Minor Works Building Contract (MW)
Minor Works Building Contract with contractor's design (MWD)
Measured Term Contract (MTC)
Prime Cost Building Contract (PCC)
Repair and Maintenance Contract (Commercial) (RM)

Subcontracts

Standard Building Sub-Contract (SBCSub/A and SBCSub/C)
Standard Building Sub-Contract with sub-contractor's design (SBCSub/D/A and SBCSub/D/C)
Design and Build Sub-Contract (DBSub/A and DBSub/C)
Major Project Sub-Contract (MPSub)
Management Works Contract (MCWK/A and MCWK/C)
Intermediate Sub-Contract (ICSub/A and ICSub/C)
Intermediate Sub-Contract with sub-contractor's design (ICSub/D/A and ICSub/D/C)
Minor Works Sub-Contract with sub-contractor's design (MWSub/D)
Short Form of Sub-Contract (ShortSub)
Sub-sub-contract (SubSub)

Other Contracts

Construction Management Appointment (CM/A)
Framework Agreement (Non-binding) (FA/N)
Framework Agreement (FA)
Pre-Construction Services Agreement (General Contractor) (PCSA)
Pre-Construction Services Agreement (Specialist) (PCSA/SP)
Consultancy Agreement (Public Sector) (CA)

The JCT also produces an adjudication agreement, a non-binding partnering charter, a large number of collateral warranties and agreements for home owners. A complete and up-to-date list of JCT documents may be obtained from its website.

With this bewildering choice of standard forms of contract, the question often arises as to which form the client should be advised to adopt; there is always a danger that in some projects, the form of contract is selected more because of the adviser's familiarity with it than because of its suitability. For example, recent Royal Institution of Chartered Surveyors (RICS) Contract Surveys[2] show that some projects let on the JCT Minor Works form were far in excess of the then recommended £125,000 limit; similarly, the Intermediate form was being used on projects in excess of the then recommended limit of £300,000. Because of the anonymity of the surveys, it is not possible to know the reasoning behind this choice of contract, but a swift inspection of the Minor Works form will show that it is a very brief document; to many, this may appear a very commendable attribute but, if problems arise on site, the contract conditions may be of little use, as the clauses cover only some of the major aspects of the construction work. They tend to give only generalised information instead of the detailed procedural information frequently found in other JCT contracts, such as the Standard Building Contract with Quantities (SBC/Q). What may be only a minor issue on, say, a £75,000 project could, if it were to occur on a £5 million project, have serious consequences, and it is necessary, therefore, to have a form of contract that is compatible with the size, complexity and type of work being undertaken.

The drawbacks to having such a multiplicity of contract forms are, firstly, the difficulty of becoming familiar with the critical obligations and liabilities of each form, and secondly, the more onerous task of identifying the subtle variations of each form. In some instances, identical contract conditions have been used within the JCT contracts, but occasionally there are slight variations to the wording of the contract conditions to reflect the specific requirements of that contract. As a result, it is possible for parties to make administrative errors because they are not familiar with the standard form of contract being used.

The following section provides a review of some of the JCT standard forms of contract in current use. The review will identify the circumstances that may be considered in the selection of an appropriate contract along with an analysis of how the contract is to be administered. Finally, each contract will be compared alongside the JCT SBC/Q to identify some of the key differences within each contract.

Minor Works Building Contract (MW and MWD)

Any clause numbers used in the following text refer to the Minor Works Building Contract with Contractor's Design (MWD).

The JCT Minor Works Building Contract (MW) is one of the smallest and simplest forms of contract produced by the JCT for commercial projects. The JCT does produce a number of very basic contracts for householders, but these contracts do not fall within the scope of this book. From a review of the Contracts in Use survey[3] commissioned by the RICS, it is possible to see that the MW contract is one of the most frequently used contracts in the UK. In the 2004 survey, it was shown that nearly 23% (by number) of the contracts captured by the survey were let on the MW form, which would seem to indicate that this is an important and widely used contract. It is certainly widely used, but its importance is diminished by the fact that it is responsible for only 2.4% of the work by value. This survey return clearly indicates that the contract is frequently used but mostly on low value projects.

Recommended Use of Minor Works

The advice currently provided by the JCT is that this contract should be used on projects where the work is of a fairly simple nature. In the Practice Notes produced for previous editions of this contract, the JCT used to advise that the contract was appropriate for projects up to a maximum contract value (in 2004, this figure was £125,000). However, in the 2005 version of the contract, the JCT has obviously decided that the advice to use the contract on works of a simple nature is a more appropriate guideline than stating a maximum contract value. Advice relating to the earlier editions of this contract also stated that the time period for the works should be short enough so that a full fluctuations provision was not required. The reason for this advice is that the MW is a fixed-price contract with an option to allow the recovery of fluctuations relating to government levies and taxes.

Interestingly, the JCT also provides recommendations for where the MW contract would not be suitable. It should not be used as a design and build contract; it is possible to require the Contractor to carry out some of the design work but the contract is not appropriate in situations where the Contractor is to be responsible for designing all the works. Furthermore, the contract is not suitable for projects where bills of quantities are required, or where work is to be carried out by named specialists or where detailed control procedures are required. These last recommendations are obviously in recognition of the fact that the MW contract was drafted on the basis that it would be used only for works of a simple nature.

In normal circumstances, the works are to be designed by the Employer (or designed on his behalf) but, where the Contractor is to design a portion of the works, this is to be clearly defined and detailed. The Employer is expected to provide the Contractor with drawings and/or specification and/or work schedules that are sufficient to allow the Contractor to assess both the quantity and the quality of the work required. Also, the Employer is expected to appoint a Contract Administrator to manage the works in accordance with the contract.

The JCT has taken a different approach from the SBC/Q in dealing with Contractor's design work in the MW contract. In the SBC/Q, there is an optional provision (using the Seventh to Tenth Recitals), which is to be used when the Contractor is required to carry out some of the design work. The same standard form may, therefore, be used for projects that have no Contractor-designed work as on those that do have an element of Contractor-designed work. The JCT did not follow this approach when drafting the MW contract and, as a result, the contract is available in two different formats. The contract may be purchased as the JCT Minor Works Building Contract or, alternatively, as the Minor Works Building Contract with Contractor's Design. The JCT has obviously decided, because of the type of client that may use the MW contract and the type of work carried out, that it would be more appropriate to have a separate contract for instances where an element of Contractor design is required.

Format

The format of the MW form is similar to the majority of the current JCT contracts. The Articles of Agreement, Recitals and Contract Particulars are placed at the beginning of the contract document, followed by the Conditions (broken down into nine sections) and finally, the Schedules. At the end of the contract, there are four pages of guidance notes, whereas other JCT contracts have guidance notes produced as a separate document. Although the MW form tends to mirror the format and structure of the SBC/Q, the content is significantly reduced. The clauses within the contract tend to be worded more briefly and simply than in the SBC/Q and a considerable amount of detail has been removed from this contract. This again reflects the fact that the contract has been developed for use on small projects of a simple nature.

Minor Works Compared with JCT SBC/Q

Articles of Agreement
One of the main differences in the articles is that there is no provision within the MW for the appointment of a quantity surveyor (QS). This is because there is no provision for a bill of quantities within the contract documents; therefore, the financial duties normally undertaken by a QS have been allocated to the Architect/Contract Administrator (A/CA). Another major omission is that there is no express procedure to provide third party rights, or for the Contractor to provide collateral warranties.

Section 1: Definitions and Interpretation
Within this section, there is confirmation that the Contracts (Rights of Third Parties) Act does not apply to this contract. A third party does not have any right to enforce a term under this contract.

Section 2: Carrying Out the Works
There are a number of omissions in the MW relating to the possession of the works and access to the works. There is no express term in the contract that allows the Employer

to delay giving possession of the site to the Contractor. The Employer has no right to use any part of the works to store goods and materials during the progress of the works, and there is no provision that would allow the Employer to take over part of the works through partial possession. Finally, there is no provision that would allow the Employer (or people engaged by the Employer) to carry out works on site while the Contractor is still in possession of the site.

In connection with carrying out the works, there is no specific requirement for the A/CA to provide the Contractor with levels and details to allow the works to be correctly set out. It would have to be implied that this information will be provided in the contract documents or through clause 2.4, where it is stated that the 'Architect/Contract Administrator shall issue any further information necessary for the proper carrying out of the Works'. There is also no requirement for the Contractor to provide a master programme to indicate how he intends to carry out the works.

Delays are a fairly common occurrence on building projects and these are normally dealt with by incorporating an extension of time clause into the contract. The procedures for dealing with a delay in the MW form are far simpler than in the SBC/Q. The Contractor still has to notify the A/CA of any apparent delay to the works, but only where this has been caused by reasons beyond his control, i.e. the Contractor does not have to give notice of delays where he is at fault. Unlike in the SBC/Q, there is no list of 'relevant events' that is to be used to identify whether an extension of time is allowable or not. As long as the Contractor can show that a delay was caused by reasons beyond his control, he should be able to claim an extension of time. Having granted the Contractor an extension of time, the A/CA does not have the power to subsequently reduce the extension if, at a later date, he omits work from the project. Similarly, after the issue of the Practical Completion Certificate, the A/CA has no power to review the current completion date, i.e. he cannot alter the completion date to an earlier or later time; therefore, under the MW form, once the A/CA has awarded an extension of time he may not subsequently alter it in the light of works being omitted or after an end-of-project review.

The MW form has a Rectification Period. The length of this period is to be stated in the Contract Particulars. If no period is stated, then by default, it will be 3 months, compared to a default period of 6 months in the SBC/Q. The A/CA may instruct the Contractor to remedy defects within the Rectification Period; there is no mention of the A/CA producing a schedule of defects as in the SBC/Q but there is nothing mentioned in the wording of the MW to prevent the A/CA from producing such a schedule. It is important to note that in the SBC/Q the A/CA has a 14-day period beyond the Rectification Period in which he may instruct the Contractor to remedy defects or to produce his schedule of defects, whereas in the MW form he must act within the Rectification Period.

Section 3: Control of the Works
Again this is a significantly cut-down version of the conditions contained within the SBC/Q. With regard to basic administration, there is no provision for the Employer to have a clerk of works on site, and there is no provision for the Employer to delegate some, or all, of his duties to an Employer's Representative. There is no express term entitling the A/CA, or his authorised personnel, to reasonable access to the works or

to the Contractor's (or sub-contractor's) workshops and premises. If, for any reason, the original A/CA ceases to be engaged for the works, there is no procedure detailing whether, or how, another A/CA is to be appointed, although it would be implied by the contract terms that another A/CA will have to be appointed (see Croudace Ltd v The London Borough of Lambeth, 1986).

As in the SBC/Q, the Contractor is required to have on site a 'competent person-in-charge'. However, this person is expected to be on site only at reasonable times, whereas the SBC/Q requires such a person to be on site at all times. Again, this subtle change in wording is a reflection of the nature of work being undertaken. On some small projects, it is not realistic to expect a Contractor to retain such a person on site throughout the whole working day.

There is a provision within the contract for the Contractor to sub-let all, or part, of the works, but there is no procedure allowing the Employer or A/CA to give the Contractor a list of preferred sub-contractors, or to name a sub-contractor. In an earlier Practice Note (Practice Note M2), it was stated that the Employer may name a sub-contractor in the tender documents, but a warning was given that there were no relevant contract conditions to help in the administration of this procedure, and there was no standard form of sub-contract available for a named sub-contractor.

Where the Contractor does decide to sub-let the works, there is only a very limited requirement as to what terms must be incorporated into the subcontract. These requirements are that any sub-contract should incorporate conditions that allow for the payment of interest on late payments made by the Contractor, and that the sub-contractor's employment is to be terminated immediately upon the termination of the Contractor's employment. By comparison, the SBC/Q requires a greater number of terms and conditions (eight in total) to be incorporated into any subcontract.

The A/CA may issue the Contractor with instructions. For the instruction to be valid, it must be in writing. In the MW form, the method of dealing with oral instructions differs from the approach used in the SBC/Q. If the A/CA issues an oral instruction, he must confirm it in writing within 2 days. The Contractor has no role to play in this situation and will have to rely upon the A/CA to comply with the contract procedure. By comparison, the SBC/Q places the onus on the Contractor to give initial written confirmation of an oral instruction within 7 days. In the 'with contractor's design' form of contract (MWD), it is only with the Contractor's consent that the A/CA may issue an instruction affecting the design of the contractor's design portion (CDP) works; the Contractor cannot unreasonably withhold his consent. However, the A/CA may alter the Employer's Requirements through a variation instruction, even if this requires the Contractor to change his design as a consequence. Unlike the SBC/Q, there is no procedure for the Contractor to query the validity of an instruction issued by the A/CA.

Variations are the key means by which the A/CA may change or develop the design during the progress of the project. Clause 3.6 details the authority of the A/CA to issue variations but, unlike the SBC/Q, there is little information on what constitutes a variation. A variation is basically defined as an instruction to make an addition to, or an omission from, the works, or a change to the works. A variation may also include an instruction to carry out the works in a specific order or time frame. The Contractor has no right of reasonable objection to this last category of variation, whereas in the SBC/Q a right of objection does exist.

Whenever a variation is issued there is almost always a cost implication. In the MW form, the value of any variations is to be agreed between the A/CA and Contractor before the work is actually carried out. In the SBC/Q, the value of variations is to be agreed between the Employer and Contractor, but there is no requirement for this to be done before the work is executed. Where the A/CA and Contractor are unable to agree the value of a variation, it becomes the responsibility of the A/CA to value the work. Very little information is given about how the variation is to be valued, apart from 'a fair and reasonable basis', using prices from the contract documents where relevant. However, the A/CA is to include any direct loss and/or expense incurred by the Contractor as a result of the variation. By contrast, the SBC/Q provides a detailed set of valuation rules for the pricing of variations and specifies that loss and/or expense should not normally be included in the valuation of a variation.

A significant role of the A/CA is to ensure that the quality of the works is in accordance with the contract documents, i.e. the Contractor is providing work to the quality and standard specified in the relevant documents. In the SBC/Q, there are specific provisions for the A/CA to open up and test the Contractor's work, and procedures that may be followed once defective works have been identified; the A/CA is also permitted to issue instructions where he considers the Contractor is not carrying out the works in a workmanlike manner. There are no comparable terms in the MW form. Although clause 2.1 identifies the Contractor's obligations to execute the works in a workmanlike manner and in accordance with the contract documents, there are no clauses identifying what the A/CA may do if the Contractor fails to comply with these standards. Again, it would have to be implied that the A/CA could take reasonable action to have the work remedied.

Finally, in the MW, there is no section dealing with the potential problem of discovering antiquities during the progress of the works. In this situation, the parties would have to rely upon any relevant statutory requirements, or reach an agreement upon how the work is to proceed.

Section 4: Payment
In the SBC/Q, it is expected that interim payments (progress payments) will be made on a monthly basis (i.e. a calendar month), whereas in the MW form, payments are to be made every 4 weeks (i.e. a lunar month). The Contractor is entitled to receive payment for the works properly executed, including the agreed value of any variations (which may include payment for loss and/or expense) and for materials on site (but there is no allowance for offsite materials). The Employer is required to pay a percentage of the progress payment. The percentage is stated in the Contract Particulars and the default figure is 95%. This is a slightly different approach to the SBC/Q, where the payment rules allow the Employer to retain an agreed percentage from payments due (e.g. normally 3% or 5%). In the SBC/Q, the amount so retained is described as retention, and the Employer is responsible for holding these monies in trust for the Contractor. By comparison, under the MW, the Employer is allowed to hold back a percentage of the progress

payment, the amount held back is not referred to in the contract conditions as retention (although retention is referred to in the heading of clause 4.3). It has no trust status and the Contractor has no right to request the Employer to hold the retained money in a separate bank account. There is no provision within the MW form for the Employer to make an advance payment, and there is no provision for the Contractor to submit his own application setting out what he considers should be included in a progress payment.

Within the payment provisions, there is reference to a 'penultimate certificate', which is to be issued 14 days after the date of practical completion. This is basically the same as a progress payment certificate, with the exception that the Employer must now pay the Contractor a higher percentage of the progress payment; the default figure is 97.5%. In effect, this certificate is releasing, to the Contractor, half of the money that has been withheld from the previous progress payments. The fact that this certificate is referred to as being a penultimate certificate must imply that the A/CA will issue no further payment certificates until the issue of the Final Certificate. In the SBC/Q, interim certificates continue to be issued after practical completion, as and when they are required.

Apart from the variations clause, there is no other provision within the MW form that would allow the Contractor to submit an application for loss and/or expense where the regular progress of the Works has been disrupted through the default of the A/CA or Employer. Under clause 3.6, the Contractor is entitled to receive payment for any loss and/or expense incurred as a result of a variation instruction or as a result of the Employer's compliance, or non-compliance, with his construction, design and management (CDM) obligations under clause 3.9. In a situation where the Contractor's progress was impeded through any other action or default of the Employer, etc., the Contractor would still have a common law right to sue for breach of contract. However, such litigation can be time-consuming and expensive. The Contractor and Employer may try to negotiate a settlement, but the contract provides no guidance to help them in this matter.

Under the MW form, the Employer does have the right to recover liquidated damages if the Contractor is late in completing the works. However, unlike the SBC/Q, there is no procedure for the A/CA to issue a non-completion certificate as a condition precedent for the Employer to recover these damages. If the Contractor is late in completing the works, the Employer may recover the damages from the Contractor as a debt, i.e. request the monies from the Contractor, or he may deduct the damages from monies due to the Contractor. If the Employer intends to recover the damages from a progress payment or Final Certificate, he must, in compliance with the Housing Grants, Construction and Regeneration Act (HCGRA), issue the appropriate withholding notice, i.e. a notice issued at least 5 days before the final date for payment. Additionally, where the Employer intends to recover the damages from the Final Certificate, he must write and inform the Contractor, no later than the date of issue of the Final Certificate, of his intention to do so.

Through the operation of clause 4.11 the contract sum may be adjusted to take into account fluctuations in government taxes and levies, and this is the default situation with this contract. However, it is possible to delete clause 4.11, in which case the project becomes a fixed-price contract, i.e. there will be no fluctuations allowed. This is obviously a reflection of the short time period envisaged for any project let under this form.

The Final Certificate procedures allow the certificate to be issued far more speedily than in the SBC/Q. The Contractor is to provide the A/CA with the necessary documentation to allow the final account to be prepared; the default period for delivering these documents is 3 months from the date of practical completion. The Final Certificate should be issued within 28 days from receipt, by the A/CA, of the Contractor's documentation (provided the Contractor has made good all notified defects and received the Certificate of Making Good). Unlike the SBC/Q, there is no advice provided as to the finality of the Final Certificate in relation to disputes that may arise after its issue, therefore it must be appreciated that the issue of the Final Certificate may not be used as conclusive evidence that certain matters have been settled. As a result, any aspect of the contract may be referred to litigation after the issue of this certificate.

Section 5: Injury, Damage and Insurance

The liability and insurance section of the MW form is similar to that of the SBC/Q, although a large amount of detail has been stripped out and certain insurances and procedures have been omitted. The Contractor's liability for damage or injury to people or property (other than to the works themselves) mirrors that of the SBC/Q. Where the works comprises all new construction, then the Contractor is required to take out the necessary insurance (clause 5.4A). Unlike in the SBC/Q, there is no optional provision for the Employer to take out insurance for all new works if he so wishes. Where the works comprises construction work within or adjoining an existing property, it is the Employer's obligation to arrange the necessary insurance (5.4B). It should be noted that the level of insurance required for new works is limited to 'specified perils' in contrast to the 'all risks' cover required by the SBC/Q. Sub-contractors should also be aware that, unlike the SBC/Q, they do not enjoy the benefit of the specified perils cover on new works.

It is possible for either the Contractor or Employer to ask for proof that the required insurances are in place. However, if either party is in default (i.e. a required insurance has not been obtained, or has lapsed, or the level of cover is insufficient), there is no express term to allow the non-defaulting party to obtain the necessary cover and to recover the associated costs.

Although the Minor Works form MWD, provides for an element of the works to be designed by the Contractor, there is no requirement for the Contractor to obtain professional indemnity insurance. If, therefore, problems arise with the Contractor's design work, there may well be no insurance in place to cover any claims made by the Employer against the Contractor. Further omissions in the insurance section, when compared with SBC/Q, are that there is no insurance procedure to cover damage to adjoining buildings where it can be proven that the Contractor was not negligent (see Gold v Patman, 1958). Also, there is no specific mention of terrorism cover and no mention of the Joint Fire Code. The lack of mention of the Joint Fire Code is understandable when it is considered that the MW form should be used only on projects of a low value. However, a few Employers do ignore the advice provided by the JCT and use the MW form on large projects that may have to comply with the Joint Fire Code.

Section 6: Termination

The termination section of the MW form is very similar to the SBC/Q. However, one slight difference between the two forms is the procedure to be followed when either party commits a 'specified default'. Under the MW form, it is permitted to terminate the Contractor's employment where the defaulting party has continued with a default for 7 days after receiving a default notice. The SBC/Q, on the other hand, requires the default to have continued for 14 days from the notice. Unlike the SBC/Q, the MW form does not contain a procedure to deal with the situation where a defaulting party ceases a default within the notification period, but repeats the default at a later period. In such a situation, the innocent party would have to go through the termination procedure all over again, whereas the SBC/Q allows for a termination notice to be issued within a reasonable time of the repeated default.

An omission from the MW form is that there is no procedure to deal with the situation where an Employer has terminated the Contractor's employment but subsequently fails to carry on and complete the works. The problem for the Contractor is that the financial statement (which sets out whether he is entitled to any further payments from the Employer, or whether he in contrast owes money to the Employer), is not prepared until a reasonable time after the completion of the works (see Tern Construction v RBS Garages, 1992).

Section 7: Settlement of Disputes

This again tends to reflect the content of the SBC/Q insofar as disputes may be referred to mediation, arbitration or adjudication. Under the HGCRA, an adjudication clause is a statutory requirement in most construction contracts, although there are exceptions. The JCT in its guidance notes identifies that, where the Employer is a 'residential occu-pier', the HGCRA is not applicable and it is not necessary to have an adjudication clause within the construction contract.

Summary

The MW Building Contract is a relatively simple form of contract and is available in two formats, i.e. with contractor's design (MWD) or without contractor's design (MW). To complement the MWD form, the JCT has produced a standard form of sub-contract to be used where a sub-contractor is responsible for some, or all, of the design work, i.e. the Minor Works Sub-Contract with sub-contractor's design (MWSub/D). However, a Contractor is not obliged to use this form of subcontract.

The MW form is similar to the SBC/Q in layout and general content, although a consid-erable amount of detail has been removed from the contract conditions. It is a lump sum contract, although it is not designed to be used with a bill of quantities. As a result, the Contractor must provide the Employer with a priced specification, or a priced work schedule or a schedule of rates. Because of its brevity, lack of detail and procedures, the MW form is adequate for simple work of a short-term nature, but should not be consid-ered for larger and more complex projects.

Intermediate Building Contract (IC and ICD)

The Intermediate Form of Contract was first published in 1984. Prior to this date, small works would be let on the MW form and anything larger would be let on the Standard Building Contract (SBC). However, in 1980, the JCT published a new edition of the SBC, which received considerable criticism from the architectural profession and employers. It was claimed that the new contract was too lengthy and complex, as well as being biased towards the contractor, so the demand grew for a simpler form of contract that would be suitable for medium-sized projects. As a consequence of this demand, the JCT introduced the Intermediate Form of Contract.

In 2005, the Intermediate Form was retitled as the Intermediate Building Contract (IC) and was made available in two versions: the basic version (referred to as the IC) and an alternative version with a provision for contractor's design (referred to as the ICD). To assist in the administration of the IC, the JCT has produced a number of complementary documents:

- Intermediate Building Contract Guide – This is a 24-page document that identifies the main differences between the 1998 and 2005 versions of this contract, and then provides a fairly detailed summary of the key conditions within the contract.
- Intermediate Sub-contract (comprising ICSub/A and ICSub/C) – This sub-contract may be used by a Contractor when engaging domestic sub-contractors. It is not mandatory for a Contractor to use these JCT sub-contract forms. However, where they are used, a Contractor can be confident that there should be no conflicts between the sub-contract and main contract forms. Similar forms are available where the sub-contractor is responsible for some of the design work (i.e. ICSub/D/A and ICSub/D/C).
- Intermediate Named Sub-contract (comprising ICSub/NAM/A and ICSub/NAM/C) – These forms are to be used where the Contractor has to engage a named sub-contractor. Where a named sub-contractor is to be appointed, the Contractor has to use the form ICSub/NAM/A and, by entering into this agreement, both the Contractor and named sub-contractor automatically agree to the contractual terms contained within ICSub/NAM/C. These forms may be used whether or not the sub-contractor has a design responsibility.
- Intermediate Named Sub-contract Tender and Agreement – This is a 27-page document that is referred to as ICSub/NAM. This document contains the forms to be completed by the contract administrator, sub-contractor and Contractor during the naming process, i.e. Invitation to Tender (ICSub/NAM/IT), Tender (ICSub/NAM/T) and Agreement (ICSub/NAM/A). The use of these forms is mandatory.
- Collateral Warranties – There are a number of standard forms produced by the JCT that may be used with the Intermediate Contract.

Recommendations for Use of Intermediate Building Contract

The general advice provided by the JCT[4] is that the IC is suitable for projects where the following criteria apply:

1. Where the building works are of a simple nature involving basic trades and skills, and where there are no complex building service installations or other specialist work of a complex nature;

2. Where the works are designed by or for the Employer;
3. Where fairly detailed contract provisions are required; and
4. Where the Employer is to provide the Contractor with drawings and other associated documents that clearly define the quality and quantity of work to be carried out.

In 2001, the JCT[5] recommended that, as long as the above criteria were met the IC would normally be appropriate for projects not exceeding £375,000 in value and where the contract period did not exceed 12 months, although a further comment was made to the effect that, in certain circumstances, the IC may be suitable for projects where these values are exceeded. To reinforce this last point, the JCT has removed any reference to time and value in its latest Practice Note and has left this aspect to the professional judgement of the Employer's advisers.

Format

The IC follows the format of the majority of the JCT contracts and is structured under the following main section headings:

Articles of Agreement
Conditions:
      Section 1: Definitions and Interpretations
      Section 2: Carrying out the Works
      Section 3: Control of the Works
      Section 4: Payment
      Section 5: Variations
      Section 6: Injury, Damage and Property
      Section 7: Assignment and Collateral Warranties
      Section 8: Termination
      Section 9: Settlement of Disputes
Schedules

From the above, it can be seen that the IC has an identical layout to the SBC/Q and it must be appreciated that, although the IC has been developed for medium-sized projects of a simple nature, the contract document is still lengthy (80-plus pages) and fairly complex. In fact, the IC is much closer to the SBC/Q in form and content than it is to the MW form. A lot of the wording used in the IC is the same as, or very similar to, the SBC/Q, although a significant amount of procedural information has been omitted. The fact that the IC and SBC/Q are similar can be an advantage in the fact that construction professionals can move from one form of contract to the other and be totally familiar with their contents. However, there are some differences in obligations and liabilities between the two forms and care should be taken to avoid confusion between the two. The following sections will review some of the key aspects of the IC and identify where the form differs from the SBC/Q.

Tendering Documentation (see 3rd and 4th Recitals)

The IC has been prepared to allow the Employer a considerable degree of flexibility in his tender documentation. The contract documents will always include contract drawings, but then the Employer is allowed a choice of documentation, as follows:

1. specification, or
2. work schedules, or
3. bills of quantities, and
4. employer's requirements for when the ICD edition is being used.

The Contractor is required to provide the Employer with a lump sum price for the works (see Article 2) and a breakdown of that price. There are two ways in which the Contractor may provide the breakdown and the Employer should identify which method is required through the Fifth Recital. For example, under Option A the Contractor is required to price whichever of the above documents has been provided by the Employer, and this will become the Priced Document. Option B will apply only where a specification has been provided as one of the tendering documents. Under Option B, the Contractor is not required to price the specification, but he is required to provide the Employer with a lump sum and either a Contract Sum Analysis, in accordance with any requirements of the Employer, or, as an alternative, the Contractor may provide a Schedule of Rates. Whichever Priced Document is provided, it will become the 'Priced Document' for the purposes of the contract. Where Option B is adopted, it is important to be aware that the ability to price variations 'accurately' will depend upon the quality of the pricing information supplied by the Contractor, therefore an Employer may be advised to request a fairly detailed contract sum analysis in preference to a schedule of rates. Finally, there is an option to request the Contractor to provide a priced activity schedule.

Intermediate Contract Compared with SBC/Q

Note: any clause references used in the following text relate to the Intermediate Building Contract with contractor's design (ICD).

Articles of Agreement

The content of the Recitals pages is very similar to the SBC/Q, although the order in which some of the information is provided has been changed. The articles are virtually identical to those found in the SBC/Q. The remaining sections, i.e. contract particulars, Third Party Rights and Attestation pages, are also very similar in layout and content to the SBC/Q, with the exception that there is no provision to provide third party rights under the Contracts (Rights of Third Parties) Act.

Section 1: Definitions and Interpretation

This section is very similar to the SBC/Q, and in many instances the terms and clauses are identical. One area of difference is that clause 1.6 clearly identifies that the Contracts (Rights of Third Parties) Act is not to apply to the IC.

Section 2: Carrying Out the Works

This section is very similar in detail to that of the SBC/Q, although the order in which a few of the clauses appear has been altered and some items have been omitted. For example, there is no requirement for the Contractor to provide a master programme. This is possibly a reflection that the IC was originally drafted for use on projects of no more than 12 months' duration and with a simple content. The procedures relating to the notification of delay and the granting of an extension of time have been slightly simplified. The Contractor must still give a notice if he feels the work is, or is likely to be, delayed but he is not required to identify whether any of the causes of the delay is a Relevant Event. Furthermore, the A/CA is required to inform the Contractor, within a reasonable time, whether or not an extension is to be granted, whereas the SBC/Q requires the A/CA to notify the Contractor within 12 weeks of receiving the required particulars. The IC does have an additional Relevant Event, i.e. A/CA instructions in relation to named sub-contractors given under clause 3.7 and/or Schedule 2. A more significant difference between the IC and SBC/Q is that the Contractor's right to request an extension of time is reduced where the works have run on beyond the completion date, or any extended completion dates (clause 2.19.2). Where the Contractor is carrying out the works beyond the completion date, the A/CA may only grant an extension of time for the following events:

1. Clause 2.20.1 variations
2. Clause 2.20.2 various instructions
3. Clause 2.20.3 deferring possession of site
4. Clause 2.20.4 inaccuracy of Approximate Quantities
5. Clause 2.20.5 rightful suspension of works by Contractor
6. Clause 2.20.6 defaults by Employer, A/CA, QS or Employer's Persons

It can be seen that all the above events are the responsibility of the Employer (or people working for the Employer); therefore, if any so-called 'neutral' events occur, (i.e. exceptionally adverse weather conditions, *force majeure*, loss or damage caused by specified perils) then no extension of time will be allowed. The logic behind this approach is that the Contractor is in breach by failing to meet the completion date and should therefore bear the risk of such delaying events that may arise after this date.

Unlike in the SBC/Q, the A/CA does not have the authority to reduce a previously granted extension of time in the light of work that has been subsequently omitted. After practical completion, the A/CA has a period of 12 weeks where he has the option of reviewing any extensions of time that have been previously granted (in the SBC/Q, the A/CA is obliged to carry out this review); he may grant further extensions of time (either as the result of reviewing a previous decision or even where the Contractor has failed to give the required notice) but again he cannot reduce any extensions of time already granted.

Section 3: Control of the Works

As in the SBC/Q, the role of administering the works on behalf of the Employer is given to the A/CA (see Article 3) plus the Employer may also engage the services of a clerk of works. Under the IC, the clerk of works acts only as an inspector; he has no right (in contrast to the SBC/Q) to issue the Contractor with directions. In accordance with most

JCT contracts, the A/CA is permitted to exercise his control through the issue of instructions. Any instructions given by the A/CA must be in writing, but there is no procedure to deal with the situation where the A/CA gives a verbal instruction. This is surprising because even the MW form places the A/CA under an obligation to provide a written confirmation, within 2 days, of any verbal instructions.

The A/CA is given the power to issue instructions through clause 3.8 and the Contractor is under an obligation to comply with them forthwith. It is important to remember that the right of the A/CA to issue instructions is determined by the contract conditions and he can do so only in circumstances allowed for by the contract. The following are some of the circumstances where an A/CA is permitted to issue instructions under the IC.

2.9	Deduction from Contract Sum for errors in setting out
2.10	Deduction from Contract Sum for defective work
2.11.1	Instructions necessary for the Contractor to complete the works
2.13.1	Inconsistency between Contract Documents
2.15.1	Divergences from Statutory Requirements
3.11	Variations
3.13	Expenditure of provisional sums
3.12	Postponement of the work
3.14	Open up work for inspection/testing
3.15	Failure of work or materials
3.16	Removal of work
3.17	The exclusion of persons from the works
4.17	The Quantity Surveyor to ascertain loss and expense
6.13.1.2	Remedial measures in response to the Joint Fire Code
Schedule 2 para. 2	Failure to enter into a sub-contract
Schedule 2 para. 4	Changing a named sub-contractor
Schedule 2 para. 5	Naming a sub-contractor against a provisional sum
Schedule 2 para. 7	Determination of named sub-contractor

It can be seen from the above that the powers of the A/CA under the IC are very similar to those in the SBC/Q, with the exception of the extra duties that come into existence where the A/CA decides to appoint a sub-contractor through the naming process. The naming process is explained below.

Under the naming process, the A/CA is able to select a specific sub-contractor to carry out part of the works, and to require the Contractor to employ that sub-contractor. A sub-contractor may be selected by the A/CA by being named in the tender documentation sent to the Contractor (procedure 1) or, during the progress of the project, by issuing an instruction against the expenditure of a provisional sum (procedure 2). A Contractor cannot make an objection to the identity of a sub-contractor named in the tender documentation, but does have the right of objection to a sub-contractor named in an instruction regarding a provisional sum. The objection must be made within 14 days from the date of the instruction.

Even though a sub-contractor has been named in the tender documentation, or in an instruction to expend a provisional sum, the A/CA may still have the work carried out by another sub-contractor. The procedure is to issue an instruction to omit the work and substitute it with a provisional sum, and thereby allow the A/CA to name the new sub-contractor in an instruction against the expenditure of that provisional sum. This procedure may only be utilised prior to the Contractor notifying the A/CA that he has entered into a sub-contract with the named person.

If an A/CA wishes to name a sub-contractor, through either procedure 1 or procedure 2, he must provide the Contractor with a full description of the work to be executed, and he must use the standard documentation provided for this process, i.e. the Tender and Agreement ICSub/NAM. This documentation is made up of three separate documents:

1. ICSub/NAM/IT Invitation to Tender
2. ICSub/NAM/T Sub-contractor's Tender
3. ICSub/NAM/A Agreement

The Contractor is to give written notification to the A/CA of the date when he entered into the sub-contract with the named person. Where a sub-contractor is named in the tender documents, the Contractor is required to enter into a sub-contract within 21 days of entering into the main contract. This obviously puts the Contractor under pressure at an early stage to finalise his programming and liaise with all his sub-contractors to agree details and dates. Likewise, if an A/CA wishes to use this procedure for naming, he must have all details finalised before the main contract tender documentation may be sent out. Where a sub-contractor is named against a provisional sum, there appears to be no stipulated time period for the Contractor to enter into a sub-contract, only a limit of 14 days in which to put forward a reasonable objection to entering into the sub-contract with the named person.

If the Contractor and proposed sub-contractor cannot reach agreement, the Contractor must promptly notify the A/CA and specify which particulars are preventing an agreement from being reached. If the A/CA is reasonably satisfied that the Contractor's notification is valid, then he is to issue an instruction, which may

- change the particulars to overcome the problem,
- omit the work, or
- omit the work and substitute a provisional sum.

The contract offers no advice as to what action the A/CA should take if he is not reasonably satisfied with the validity of the Contractor's notification. Once an agreement has been reached between the Contractor and sub-contractor, the Contractor is to inform the A/CA of this fact and confirm the date the agreement was concluded.

Whichever naming procedure is adopted by the A/CA, the Contractor is obliged to pay the named sub-contractor the amount quoted in NAM/T. Where the sub-contractor is named in the tender documents, the Contractor is obliged to price the work described in the tender documents and will receive payment for this from the Employer via interim valuations. The payment procedure is not quite so simple, however, where a sub-contractor has been named post contract (i.e. as the result of an instruction to expend a provisional sum). In this instance, the amount to be paid to the Contractor will have to be agreed between the Contractor and Employer or, where there is no agreement, the amount will be determined by the QS (clause 5.2). No advice is given on how this amount is to be determined. Whichever valuation method is used, the named sub-contractor will not be affected as they are guaranteed payment in accordance with their tender NAM/T.

The Contractor is to advise the A/CA, as soon as is reasonably practicable, of any events that may lead to the termination of the named sub-contractor's employment. If a named sub-contractor's employment is terminated, then the Contractor must inform the A/CA, stating the relevant circumstances. Where a sub-contractor's employment is terminated, the A/CA must issue an instruction:

- naming a replacement sub-contractor, or
- for the contractor to make his own arrangements for completion (i.e. the contractor may complete the work himself or sub-let the work), or
- omitting the remaining work.

The above has been only a brief review of the 'named sub-contractor'; the full procedural details may be found in Schedule 2 of the IC.

If the Contractor wishes to sub-let the design for his CDP work, he must first get the permission of the A/CA (not the Employer as in SBC/Q).

There is no procedure to deal with the discovery of antiquities.

There is a completely different procedure for dealing with defective works. In the IC, upon the discovery of work or materials that are defective, the obligation is on the Contractor to provide written proposals of what action he will take to show that there is no other similar defective work elsewhere. If the Contractor fails to provide this information within 7 days, or his proposals are considered inadequate, the A/CA may issue the Contractor with instructions to carry out further tests as necessary. If a defect is considered to be so serious that the A/CA is not prepared to wait 7 days for the Contractor's proposals, then again the A/CA may issue instructions for further tests to be carried out. All these tests are at the Contractor's expense. If the A/CA does issue the Contractor with an instruction to carry out further tests, etc., the Contractor has the right to challenge the instruction if he considers it to be unreasonable. If the Contractor is not satisfied with the A/CA's response to his challenge, the issue may be resolved through the use of one of the dispute resolution procedures. Finally, there is no procedure, during the progress of the works, for the A/CA to instruct the Contractor that an item of defective work may remain and for the contract sum to be subsequently adjusted, although such an instruction may be issued during the Rectification Period (clause 2.30).

Section 4: Payment Provisions

The payment provisions under the IC are very much the same as under the SBC/Q. The Contractor is entitled to receive payments on account for the work executed and for materials on site. The payment is usually made on a monthly basis as a result of interim valuations prepared by the QS, or based on an application submitted by the Contractor. The Employer is to pay the Contractor within 14 days of the date of the Interim Certificate. However, the Employer is generally only required to pay a percentage of the value of work and materials (4.7.1), although some items have to be paid in full, e.g.

insurance premiums, statutory fees, loss and/or expense claims (4.2.2). The default percentage to be paid by the Employer is 95%, although it is possible to change this in the Contract Particulars.

The 5% that has not been paid by the Employer is released in the following manner:

1. 2.5% within 28 days of date of Practical Completion
2. 2.5% after 28 days from date of Final Certificate

The money that is withheld by the Employer is not referred to as retention, as in SBC/Q, although it is treated in a similar manner. Where the Employer is not a local authority, it is stated that the amounts not paid in interim payments shall be held by the Employer as 'fiduciary trustee', which means the money is held in trust by the Employer for the benefit of the Contractor. However, unlike under the SBC/Q, there is no provision for the Contractor to request that this money be held in a separate bank account as a safeguard in case the Employer gets into financial difficulties. Also note that the final 2.5% is released after the issue of the Final Certificate and not at the end of the Rectification Period or issue of the Certificate of Making Good, whichever occurs last, as under the SBC/Q.

Through clause 4.15 the default position of the IC is that fluctuations are limited to changes in taxes and levies, although it is possible to delete this option in the Contract Particulars and as a result there would be no fluctuations allowed. Interestingly, named sub-contractors may be required by the Employer to operate under the same conditions as the main contract (i.e. no fluctuations or limited fluctuations) but they may also be allowed to claim full fluctuations under the formula method (i.e. a method of recovering fluctuations that is not available to the Contractor). In order to determine what fluctuations apply to a named sub-contractor, it will be necessary to look at section T3 of the sub-contract tender (ICSub/NAM/T).

The rights of a Contractor to request additional payment to cover loss and/or expense resulting from a disruption to the regular progress of the works are very similar to the SBC/Q.

Section 5: Variations

It is usually essential in any building project to make allowance for variations to the design and quality of materials. Again the basis for issuing and valuing variations is very similar to the SBC/Q. It is initially expected that the Employer and Contractor will agree the value of any variations; if they cannot agree, the task will pass to the QS (unless the Employer and Contractor appoint another party) to value the variation in accordance with the 'valuation rules'. The valuation rules are very much reliant upon the Priced Document, which may be a priced bill of quantities, priced work schedules, priced specification or a contract sum analysis or a schedule of rates. Variations to CDP work are to be largely based upon the CDP Analysis.

Sections 6 to 9

The final sections of the IC deal with insurances, assignment and collateral warranties, termination and settlement of disputes and generally follow the procedures contained in the SBC/Q. The one major difference is that there is no reference to giving third party rights through the Contracts (Rights of Third Parties) Act, as this legislation is specifically excluded from the IC.

Summary

The IC is designed for use on medium-sized projects of a fairly straightforward nature. The Employer must be in a position to provide the Contractor with fairly detailed design documentation to enable the Contractor to provide a lump sum tender, although it is possible to have the Contractor design a portion of the works through the use of the contractor-designed version of the Intermediate Contract (ICD). The Employer is also expected to engage the services of an architect (or contract administrator) and a QS, and if he wishes he may also employ a clerk of works. The layout and content of the IC is very similar to the SBC/Q, although a few of the clauses have been simplified and some items have been omitted altogether. Possibly the most notable difference between the two forms is the introduction of the named sub-contractor procedures contained within the IC.

Design and Build Contract (DB)

Since the 1980s, there has been a steady growth in the use of the Design and Build procurement approach, and in 2004 it was estimated that nearly 43% of building work (by value) was being undertaken by this method.[6] The DB contract is suitable for projects where the client wants a contractor to take on the responsibility for both the design and construction work, for an agreed lump sum. The format and the conditions of contract are very similar (and sometimes identical) to the SBC/Q, especially as the SBC/Q now incorporates conditions that deal with CDP Work. Because there is such a similarity between the DB and SBC/Q forms of contract it is important to be aware of where the two forms differ.

Contract Documents

The contract documents comprise the agreement and conditions (i.e. the DB form of contract), the employer's requirements, the contractor's proposals and a contract sum analysis. There is no reference to drawings being contract documents, although they will obviously be contained within the Employer's Requirements and the Contractor's Proposals. There is no bill of quantities, although they may be introduced through the use of the supplemental provisions, which are to be found in Schedule 2. There is no mention of an A/CA, clerk of works or QS, although the Employer is expected to appoint an Employer's Agent (see Article 3). This person may be a construction professional (e.g. architect, engineer, QS or project manager) or an individual from the Employer's organisation. Although this person is named in Article 3, the Employer may replace them at any time. The Employer should notify the Contractor of any new nomination, but does not need to have the Contractor's consent. The Employer's Agent has full authority to act on behalf of the Employer under these conditions. If the Employer wants the Employer's Agent to have only limited authority under these conditions, he must write to the Contractor and notify him of the limitations. As there is no A/CA in the DB contract, it becomes the responsibility of the Employer (or Employer's Agent) to issue certificates and instructions, and to grant extensions of time.

Carrying Out the Works

The obligations of the Contractor for carrying out the works are basically the same as under the SBC/Q, where CDP work has been included; i.e. to carry out and complete the works, complete the design works and select materials where they are not specified in the Employer's Requirements or Contractor's Proposals and give all necessary statutory notices. Two additional clauses are to be found in this section of the DB: firstly, there may be a requirement for the Contractor to provide samples of the standard of workmanship or quality of goods to be provided (clause 2.2.3) and secondly, there is a requirement that the Employer must define the site boundaries (clause 2.9). The Contractor's right to request an extension of time is very similar to the SBC/Q. The relevant events that allow the granting of an extension of time are very similar, although the DB does have an additional relevant event where there has been a delay in receiving permission or approvals from a statutory body, and which the Contractor has taken reasonable steps to reduce (e.g. this could relate to problems with gaining planning permission).

Again there is very little difference between the DB and SBC/Q with regard to the control of the works (Section 3), although under the DB contract, the Employer has no right to issue an instruction excluding one or more of the Contractor's operatives from the site.

Payment

With regard to interim payments under the DB contract, there are two methods that may be used to value the work carried out by the Contractor: Alternative A and Alternative B. The Contract Particulars should identify which option is to be used; the default position is that Alternative B is to be used. Under Alternative A, the Contractor is entitled to be paid on a stage payment basis. A number of work stages will be identified in the Contract Particulars and priced on a cumulative basis. As the stages are completed, the Contractor is to submit an application for payment for the cumulative value of the works (taken from the Contract Particulars) at that stage. Under Alternative B, the Contractor is to make an application for payment on the dates set out in the Contract Particulars. The payment is based upon the value of work and design work executed by the Contractor by that date. Alternative B is basically the procedure used in the SBC/Q, with the exception that the value of work is assessed on the contract sum analysis submitted by the Contractor, instead of on a bill of quantities. Regardless of whether Alternative A or B is used, the contractor may also claim the cost of Changes (variations), listed offsite materials and fluctuations. Materials on site may only be claimed under Alternative B. The interim payments are based on the Contractor's applications, and the Employer must pay within 14 days from the date the application was received. The Employer is able to deduct retention from payments from the Contractor, but there is no alternative provision to allow a retention bond as a replacement for retention.

The Contractor is the party responsible for preparing the final account and a final statement. These documents should be submitted to the Employer within 3 months of practical completion. The final account is the Contractor's assessment of the total cost of the works and the final statement is like the Final Certificate under the SBC/Q; it states the amount of money payable to the Employer or Contractor in settlement of the final account. If the Contractor fails to produce the final account and final statement within the required time frame, the Employer may take on the responsibility for producing these documents. However, this is an option, not an obligation. If the Employer intends to take on this task, he must have allowed the Contractor the 3-month period to produce the documents and then give the Contractor a notice to the effect that, if the Contractor fails to produce the documents within the next 2 months, the Employer may himself prepare the documents. Even if an Employer issues this notice, he is under no obligation to prepare the documents; it is still his option. Regardless of which party prepares the final account and financial statement, the documents are conclusive evidence that the amounts are correct, unless they are challenged within the appropriate time.

A Contractor's right to claim for loss and/or expense is similar to the procedures under the SBC/Q, although there is an additional event in DB that allows the Contractor to submit a claim. This additional event relates to delays in obtaining permissions or approvals in relation to development control, e.g. planning permissions.

The section dealing with variations is similar to SBC/Q. One of the most obvious differences is that variations are referred to as 'Changes' in the DB contract. The value of any Changes is to be agreed between the Employer and Contractor. If they cannot agree on a value, the next step will normally be for the work to be valued in accordance with the valuation rules as in SBC/Q, but as in this instance there is no QS to apply the rules, it is once more down to the Employer and Contractor to reach an agreement by using the valuation rules. Although the wording of the valuation rules differs slightly between the DB and the SBC/Q, the general principles to be used in valuing the work are the same. The biggest drawback to using the valuation rules under the DB is the lack of detailed cost information. In the SBC/Q, the bill of quantities provides a detailed breakdown of the Contractor's costings, whereas in the DB contract the bill of quantities is replaced by a contract sum analysis, which normally provides only a very broad breakdown of the project costs and may be of very limited use when trying to apply the valuation rules.

Schedule 2

There are a number of schedules at the back of DB. Possibly one of the most significant ones is Schedule 2, which introduces a number of additional contract conditions that can impact substantially on the administration of the project. The use of Schedule 2 is optional and reference will have to be made to the Contract Particulars to see whether the Schedule applies or not.

The first paragraph of the Schedule is an optional provision and again the Contract Particulars will identify whether or not it applies. If the paragraph does apply, the Contractor is required to appoint a site manager to act in place of the competent person in charge referred to in clause 3.2. The person selected to be site manager must be approved by the Employer and be appointed (by the Contractor) before work starts on site. Having made the appointment, the Contractor cannot remove or replace the site manager without the Employer's consent, which cannot be unreasonably delayed or withheld. The site manager is to attend meetings called by the Employer when reasonably required to do so, and must keep accurate site records that comply with any specific requirements of the Employer, as set out in the Employer's Requirements.

The second paragraph of the Schedule allows the introduction of 'Named Sub-contractors'. Through this process an Employer may use the Employer's Requirements to name a person whom the Contractor is required to engage as a sub-contractor. The naming procedure in the DB is different from the IC to the extent that the JCT has not produced any standard documentation to assist with the procedure; there are no standard forms of tender and agreement.

Where the Employer has named a person in the Employer's Requirements, the Contractor is normally required to enter into a sub-contract with that person as soon as possible after entering into the main contract. On some occasions, it may not be possible for the Contractor to enter into a sub-contract with the named person, e.g. there may be basic disagreements about the service the named person is able to provide, the timing of the service, or problems with insurance or liability. If this situation arises, the

Contractor must inform the Employer. If the Employer accepts that the Contractor has a genuine problem with the named person, he must do one of the following:

1. Resolve the problem by issuing a change instruction that would then allow the Contractor to enter into a sub-contract, or
2. Issue a change instruction to omit the named sub-contract work and follow that up with an instruction that details how the work is to be dealt with.

If the Employer chooses to omit the named sub-contract work, he may not subsequently instruct that the work be carried out by another named person, but he may instruct the Contractor to find another person to carry out the work. Alternatively, the Employer could arrange to have the work carried out separately from the main contract, as allowed under clause 2.6.

Once a named sub-contractor is appointed, the Contractor should not terminate his employment without the Employer's consent. Where a named sub-contractor's employment has been terminated, the Contractor becomes responsible for completing the remaining sub-contract works. The Contractor will be paid for these works as though they were a 'Change', unless he failed to obtain the Employer's consent to terminate the named sub-contractor's employment, or the termination was as a result of a default of the Contractor. The Contractor must inform the Employer of any monies he is able to recover from the named sub-contractor as a result of the termination, and these monies are to be taken into account when valuing the Change.

Paragraph 3 gives the Employer the option of using bills of quantities to describe the work in the Employer's Requirements. This could be a realistic option only where the works have been substantially designed by, or on behalf of, the Employer before the Contractor is asked to submit his proposals, e.g. a novated design and build project. If the Employer wishes to use bills of quantities, the Employer's Requirements must state what measurement rules were used to prepare the bills. If there are any mistakes in the bills, they are to be corrected and treated as a Change to the Employer's Requirements; therefore, any errors will be at the Employer's expense. The advantage of using bills of quantities is that they provide a much more detailed breakdown of the Contractor's costs than would a contract sum analysis. They can therefore assist the Employer in agreeing the value of Changes and assessing the value of formula fluctuations.

Paragraph 4 introduces a procedure that is similar to the SBC/Q Schedule 2 Quotation. The procedure is triggered off by the Employer issuing an instruction that would, in the opinion of the Employer or Contractor, require:

1. the valuation of a Change or provisional sum work, and/or
2. an adjustment of time for a Relevant Event, and/or
3. an assessment of loss and/or expense under clause 4.19.

In the event of one or more of the above situations arising, the Contractor is not to comply with the instruction and must, within 14 days, or other agreed time, provide the Employer with the following estimates, where relevant:

- the required adjustment to the contract sum for complying with the instruction;
- what additional resources may be required;
- a method statement;
- any extension of time that may be required; and
- the amount of any direct loss and/or expense that may be incurred.

The Contractor and Employer should try and agree the estimates within 10 days of receipt, whereupon they become binding on both parties. If agreement cannot be reached, the Employer may withdraw the instruction, or instruct that it is to be complied with and dealt with under the normal contract conditions. If the Contractor fails to comply with the requirements of paragraph 4, then the instruction will be dealt with under the normal contract conditions, but the Contractor will not be entitled to any payment for the instruction until the issue of the Final Statement.

It is important to note that there are instances when this 'Paragraph 4' procedure may be ignored. For example, an Employer may state in the instruction (or within 14 days of issuing the instruction) that the Contractor is not required to provide estimates. Also, within 10 days of the issue of the instruction, the Contractor may put forward an objection to complying with all or some of the procedure. The Contractor's objection has to be reasonable.

Paragraph 5 modifies the contract conditions relating to the assessment and payment of direct loss and/or expense. Clause 4.19 entitles the Contractor to receive payment for loss and/or expense on the basis that he informs the Employer of the disruption to the progress of the works and provides information to support his application. As the loss and/or expense is assessed, it will be paid to the Contractor through interim payments. Under paragraph 5, the Contractor is required to provide the Employer with estimates of his loss and expense at the same time as his applications for an interim payment; therefore, the Contractor is keeping the Employer informed on a regular basis of his estimated loss and expense. Within 21 days of receiving the estimate, the Employer may, if he wishes, ask the Contractor for more information and details to support the estimate, but within the same time period the Employer must give the Contractor one of the following notices to the effect that:

1. the Employer accepts the estimate, or
2. the Employer wishes to negotiate the estimate, and if agreement cannot be reached it is to be referred to adjudication, or
3. clause 4.19 is to be used to deal with the claim in place of paragraph 5.

Where the estimates are accepted or agreed, they will determine how much the Contractor receives; they cannot be adjusted at a later stage. If the Contractor fails to provide the estimates as required, the loss and expense will be assessed under clause 4.19 but the Contractor will not be entitled to payment until the issue of the Final Statement.

Summary

The DB contract is a lump sum contract and there are three key contract documents to be used in conjunction with it. Firstly, the Employer's Requirements, which is a document similar to a client's brief, identifying what the Employer wants from his completed building. This document may be prepared in some considerable detail or it may provide just basic requirements, leaving the Contractor to flesh out the detail. Secondly, in response to the Employer's Requirements, the Contractor produces and submits his own proposals (the Contractor's Proposals); this document illustrates how the Contractor intends to design and produce a building to meet the Employer's Requirements. Thirdly, to assist with the financial administration of the project, the Contractor will provide a breakdown of his lump sum tender by producing a Contract Sum Analysis.

The DB contract is a complex and fairly lengthy document containing just over 100 pages. It is very similar in size and complexity to the SBC/Q, and there is a high degree of commonality between the contract conditions of the two forms. There is a significant difference with regard to the administration of a DB project in that there is no A/CA to administer and control the works and no QS to assist with the financial aspects of the works. The responsibility of administering the project rests with the Employer and Contractor, although the Employer is able to delegate this role to an Employer's Agent.

As previously mentioned, many of the contract conditions contained within the DB contract are identical to those found in the SBC/Q, especially now that the SBC/Q contains procedures for CDP works. Some of the differences that do exist between the two forms have been identified in the above text, but perhaps some of the most significant differences lie within the optional clauses contained within Schedule 2 of the DB contract.

Management Building Contract (MC)

The Management Building Contract (MC) is predominantly designed for large-scale and possibly complex projects where the Employer would like an early start on site at a time when the detailed design is still to be agreed. Under the management contracting approach, the Employer engages a contractor (the management contractor) to supply the site facilities and to perform the site management role. The Management Contractor will in turn engage sub-contract organisations (works contractors) to carry out the construction work. The Employer will engage a number of professional advisers to assist with the design and cost implications of the project, although some of the design work can be passed on to the works contractors. Where the Employer uses the MC contract, he is obliged to appoint an A/CA and a QS (Articles 3 and 4); the other members of the Consultant Team are identified in Article 5, or may be notified to the Management Contractor at a later date. The Management Contractor's key personnel are identified in the Contract Particulars.

Format

The format and layout of the MC reflects the current style adopted by the JCT and the document is broken down into the following sections:

Articles of Agreement
Conditions:
    Section 1: Definitions and Interpretations
    Section 2: Carrying out the Project
    Section 3: Control of the Project
    Section 4: Payment
    Section 5: Works Contracts
    Section 6: Injury, Damage and Insurance
    Section 7: Assignment, Third Party Rights and Collateral Warranties
    Section 8: Termination
    Section 9: Settlement of Disputes
Schedules
Annexes:
    Annexe A: Site Facilities
    Annexe B: Services

Articles of Agreement

A number of items included in Part 1 of the Contract Particulars are not normally found in JCT contracts. With reference to clause 1.1, the Contractor is required to list the names of his key personnel engaged on the project and identify their role. The accompanying footnote explains that it is important to indicate which key personnel are site-based and which are off site, the reason being that the financing of the site-based personnel is dealt with as part of the prime cost of the project, whereas the cost of the offsite personnel is covered by the management fee. A Review Date is to be inserted into the Contract Particulars (see clause 2.1.2); the Consultant Team and Contractor are expected to agree the basic pre-construction details by this date. Details of the Prime Cost are to be inserted (see clause 4.1.1); these details are only provisional at this stage and are supplied as an aid to negotiations during the pre-construction phase. Details are also to be provided to show how the management fee is to be calculated and, finally, details of the pre-construction services are to be supplied by the Management Contractor.

Because of the uncertain nature of a management construction procurement approach, it is not always possible to set out firm details in the contract documents during the pre-construction phase. Once this phase has been finalised therefore, it may be necessary to firm up or alter some of the previous provisional details in Part 1 of the Contract Particulars. This may be achieved through the use of Part 3 of the Contract Particulars, i.e. the Supplemental Particulars. Here the parties can set down the Project Total Cost Plan, confirm the dates of possession and completion, provide details of the Prime Cost that may have been agreed, confirm how the management fee is to be calculated and set out details of the services to be provided by the Management Contractor during the construction period.

Section 1: Definitions and Interpretations

The MC form of contract is similar (or identical) to the SBC/Q in many areas but, because of the management role of the Contractor, there are also a number of substantial and minor differences between the two forms. Some of the minor differences relate to the wording used in the MC contract; e.g. the main works are referred to as the Project, sub-contractors are referred to as Works Contractors and the sub-contract works are referred to as the Works. The MC also introduces some terms and concepts that may be unfamiliar to users of the SBC/Q, e.g. Acceleration Quotation, Consultant Team, Construction Period Fee, Pre-construction Fee, Management Fee, Site Facilities, Review Date, Notice to Proceed and Project Cost Plan. All these terms are identified within Section 1, along with a brief explanation of how the terms are to be used within the MC.

Section 2: Carrying Out the Project

In this section, the MC generally follows the layout and wording of the SBC/Q, with the exception of the earlier clauses that reflect the different role taken on by a Management Contractor. For example, before the Management Contractor can make a start on site he requires a 'Notice to Proceed'. To obtain this notice, the Management Contractor has to comply with certain obligations; during the pre-construction period he must co-operate with the Consultant Team and provide the Pre-construction Services identified in Annexe B. Furthermore, on or before the 'Review Date' (a date stated in Part 1 of the Contract Particulars), the Management Contractor is required to liaise with the Consultant Team so that agreement may be reached on the following details:

1. The Project Cost Plan that is to be prepared by the QS in collaboration with the consultant team and management contractor;
2. The date of possession, if it is to be different from the one provisionally identified in Part 1 of the Contract Particulars;
3. The date for the completion of the project;
4. The specific details regarding the supply of the site facilities and construction period services;
5. The final project specification and drawings, where necessary;
6. The prime cost (i.e. the cost of the project calculated in accordance with Schedule 1); and

7.   The supplemental particulars to be found in Part 3 of the Contract Particulars, which is where certain provisional and contract detail is finally set out in the agreement.

Following this consultation period, the A/CA should notify the Employer of the proposed terms and advise whether it is practicable to start the construction work. If the A/CA fails to notify the Employer within the required time frame, the Employer may be given a notice by the Management Contractor warning that his employment will be terminated unless the A/CA gives a proper notice. Where the A/CA has given a proper notice, and the Employer accepts the proposed terms, he should inform the Management Contractor to set the construction work in motion, i.e. give the notice to proceed. If the Employer fails to properly give the notice to proceed, the Management Contractor has the right to request the Employer to issue either the notice to proceed or a notice terminating the Management Contractor's employment. If the employer fails to respond to this request within the due time, the Main Contractor's employment is deemed to be terminated. Under normal circumstances, the Management Contractor would still receive his Pre-construction Period Fee.

Having received a notice to proceed, the Management Contractor must ensure that the project is carried out in a workmanlike manner and in accordance with the contract documents, the contract documents being the project drawings, project specification, project cost plan, as well as the agreement, conditions and Annexes A and B of the MC form. Furthermore, the Management Contractor is also required to carry out the project in an economic and 'expeditious' manner. This last requirement is an attempt to ensure that the Management Contractor undertakes his management role effectively. The Management Contractor is also required to fulfil further management tasks during the construction period, e.g.:

- Collaborate with the Consultant Team, prepare project programmes as necessary and provide the construction period services listed in Part 2 of Annex B;
- Enter into works contracts in a timely manner to allow the project to be completed on time;
- Ensure the required site facilities are provided;
- Supervise and monitor the works contractors;
- Provide continual project supervision and assist with the management and organisation of the project; and
- Keep records, as required, to allow the QS to be able to assess the prime cost of the project.

Section 3: Control of the Project
This section sets out the administrative detail and procedures for the control of the project. Although some of the detail contained within the MC is the same as in the SBC/Q, there are procedures within the MC form that differ from the SBC/Q. Some of the differences to be found in the MC form are as follows:

- The A/CA and QS have the right to inspect the Management Contractor's project documentation to assist in both the running of the project and the assessment of payments due to the Management Contractor.

- The Management Contractor must ensure that his key personnel are normally available and fulfilling their roles on the project and that he maintains the necessary management personnel for the supervision of the project and the site works.
- The selection of works contractors is to be agreed between the Management Contractor and A/CA. Normally, a works contractor is required to be willing to enter into a current and unamended form of Management Works Contract Agreement, a Management Works Contractor/Employer Agreement (a collateral warranty) and into third party rights or collateral warranties where required. It is important for the Management Contractor to appreciate that he is liable to the Employer, in most instances, for the performance of the works contractors. If the Management Contractor wishes to sub-let some of his pre-construction or construction period management services, he must get prior approval from the Employer, or from the A/CA acting on behalf of the Employer.
- An innovation of the MC is that the A/CA may issue an instruction requesting an 'Acceleration Quotation', the procedures for which are set out in Schedule 6. There are some slight similarities between an Acceleration Quotation and the Schedule 2 Quotation to be found in SBC/Q but the main purpose of an acceleration quote is to obtain an earlier completion date for the project, or a section, or to try and reduce an extension of time that is being currently assessed, or to alter the sequencing or timing of any work on the project. The Management Contractor can decline to give a quotation if he has a reasonable objection. If not, he must inform the A/CA of the proposed revised completion date, and of the lump sum required by each works contractor affected by the instruction. In some instances, he may state that it is impracticable to calculate a lump sum and that the work will have to be assessed under the normal contract conditions. If the Employer accepts the quotation, the A/CA is to issue a Confirmed Acceptance.

Section 4: Payment

The payment procedures of the MC tend to mirror the procedures contained within the SBC/Q, although there is a significant difference in the items for which a contractor is entitled to receive payment, e.g. the Management Contractor is entitled to receive payment for his management fee, which is broken down into a pre-construction fee and a construction period fee. The pre-construction fee may be a fixed fee or a variable fee, depending on the contract documentation, and will be paid through interim certificates issued before the date of possession, at the times set out in the Contract Particulars. Similarly, the construction period fee may be a fixed fee or an adjustable fee. Where a fixed fee applies, and also Schedule 2 applies (an optional schedule), the fee is to be adjusted if the prime cost of the work varies by more than 5% (or other agreed percentage) from the project cost plan. The fee is adjusted in accordance with the formula set out in Schedule 2, and the end result is that the fee will be reduced if the cost plan is exceeded, or increased if the prime cost is less than the cost plan. During the construction period, the A/CA will issue interim certificates based on interim valuations prepared by the QS. The main items to be included in the valuation are: amounts due to the works contractors, Management Contractor's staff and workforce costs, materials, plant and services provided by the Management Contractor, the pre-construction fee and an appropriate portion of the construction period fee (up to a maximum of 97%

of the fee). With each interim certificate issued the A/CA must inform the Management Contractor of the amount of money that has been included for each works contractor. Furthermore, a works contractor has the right to request the A/CA to send him a statement identifying the amount of money included for him in an interim certificate and with a breakdown to indicate how this sum was calculated. The Employer is entitled to withhold retention from the prime cost of the project (the default figure is 3%). At the issue of each interim certificate, a statement has to be prepared to show what retention is being withheld from the Management Contractor and from each works contractor. Each works contractor is to be separately informed of the amount of retention withheld from their payments. It is possible for a works contractor to provide a retention bond as an alternative to having retention deducted by the Employer. Under the MC, it is also possible for a works contractor to receive an early final payment, i.e. where a works contractor has completed his work package he is able to request a full payment of his works without having to wait for the issue of the Final Certificate for the project. The decision whether or not a works contractor is to receive an early final payment rests with the Employer, or with the A/CA acting on behalf of the Employer. If a works contractor is to be successful in his request, he must be able to demonstrate that he has made adequate arrangements with the Management Contractor for dealing with any latent defects that may be present in the works.

Section 5: Works Contracts

In this section, the MC deals with the administration issues relating to the works contractors, such as the procedures to be followed where a works contractor is in breach of his contract, or where a works contractor claims that the Management Contractor is in breach of contract, claims for loss and/or expense and termination of the works contractor's employment. Under the SBC/Q, the Employer and A/CA have virtually no involvement with the administration of sub-contractors. However, under the MC, they become far more actively involved in dealing with issues relating to the works contractors and to agreeing claims and payments.

Section 6: Injury, Damage and Insurance

The insurance provisions of the MC are very similar to the SBC/Q. For example, the Management Contractor has the same liability for injury to property and persons, and there are the same options for the project works and site materials to be insured by either the Management Contractor or Employer. Similarly, the works contractors receive some recognition under the project works insurance. There is also an option to require the Management Contractor to take out professional indemnity insurance and there are procedures for dealing with the Joint Fire Code. The professional indemnity insurance taken out by a Management Contractor is to provide cover for the professional services he supplies to the Employer, and not for any design liability. The reason for this is that the Management Contractor, in his management role, does not take on any design responsibilities. Any contractor-designed work will be undertaken by the works contractor who is therefore required to provide the appropriate professional indemnity insurance.

Section 7: Assignment, Third Party Rights and Collateral Warranties

The procedures for assignment, third party rights and collateral warranties differ very little from the SBC/Q, with the exception that works contractors can be required to provide rights under the Contracts (Rights of Third Parties) Act, as well as provide collateral warranties. Under the SBC/Q, there is provision to require sub-contractors only to provide collateral warranties. Under the MC, all works contractors have to be engaged under the JCT standard form for works contractors, therefore it is fairly straightforward to incorporate third party rights into the works contractor's documentation.

Section 8: Termination

The termination rights of the Management Contractor and Employer are basically the same as under the SBC/Q, although under the MC the Employer does possess an additional right of termination through clause 8.4. Under this clause, the Employer may terminate the Management Contractor's employment for no reason at all. This is not a right of termination usually found in JCT contracts, and is perhaps a reflection of the role undertaken by the Management Contractor under MC. The Management Contractor plays an advisory and supervisory role on the project and, if there is a breakdown of trust or communication with the Employer, the Employer may feel he is unable to continue working with the current Management Contractor. With reference to termination, there is normally a requirement that any notice of termination should not be seen to be unreasonable or vexatious (clause 8.2.1) but this does not apply where the Employer issues a notice under clause 8.4. However, in such a situation the Management Contractor would receive payment for works and services provided prior to the termination. The actual amount for which a Management Contractor may claim after a clause 8.4 termination is dependent upon when the termination notice is issued. If the termination takes place before the notice to proceed, the Management Contractor is entitled to receive a proportion of the Pre-construction Fee. The amount received should reflect the level of assistance that the Management Contractor has provided to the Consultant Team in accordance with the requirements of clause 2.1. If the termination takes place after the Notice to Proceed the Management Contractor is entitled to receive, amongst other items, the Pre-construction Fee, an appropriate portion of the construction period fee, the reasonable cost of removal from the site and any loss and/or damage caused by the termination (clause 8.13.3.5). This could include a claim for the loss of profit that the Management Contractor expected to achieve on this project.

Section 9: Disputes

The way in which disputes are to be settled is virtually the same as under the conditions contained within SBC/Q.

Summary

The management contract provides the Employer with the services of a Management Contractor. The role of the Management Contractor is to assist the Consultant Team in their administrative duties, to provide site services in support of the construction work and to engage works contractors to carry out the construction work. The design work for the project is undertaken by the Consultant Team or by the works contractors; the Management Contractor has no design liability.

There are two distinct phases to the project: the pre-construction period and the construction period. If there is inadequate progress made during the pre-construction period, there is a provision that allows the Management Contractor's employment to be terminated. If satisfactory progress is made during the pre-construction period, a notice to proceed will be issued, allowing the project to progress into the construction period.

The works contractors engaged by the Management Contractor will normally work on a fixed-price basis, although there is an option that allows them to work on a remeasurement basis. The work carried out by the works contractors, along with the Management Contractor's site services, comprises the Prime Cost of the project. The Employer pays the works contractors via the Management Contractor. One innovation to be found in the MC contract is the introduction of an Acceleration Quotation, a means by which the Management Contractor can be requested to advise the Employer of the financial cost of completing the works before the agreed completion date, or of reducing an extension of time that has been applied for, or of altering the sequence of the works.

Constructing Excellence (CE)

The Constructing Excellence Contract (CE) is one of the newer forms of contract produced by the JCT and was first published in 2006. Its format and style are significantly different from the more traditional JCT forms of contract and it also introduces a number of innovative ideas. The CE form of contract is actually based on a previously published contract (Be Collaborative Contract), whose principal authors were solicitors Martin Howe and Giles Dixon, and this possibly explains the different approach adopted by the JCT when producing the CE. The JCT has produced three documents for use with this procurement approach, i.e.:

- Constructing Excellence Contract – CE
- Project Team Agreement – CE/P
- Constructing Excellence Contract Guide – CE/G

The CE is very much a multi-purpose form of contract. It may be used for the procurement of both construction works and construction-related services; therefore, contractors and construction professionals, such as architects, engineers, designers and QSs, can all be appointed under an identical form of contract. Also, in relation to the construction works, it is anticipated that the same CE form of contract will be used to engage organisations further down the supply chain, such as sub-contractors and sub-sub-contractors, as well as material and plant suppliers. One of the key aspects of the CE is that it is a form of contract to be used where the parties wish to work on a collaborative basis and it encourages the use of integrated teamwork throughout the supply chain. As a consequence, the CE is a suitable form of contract where a partnering approach is to be employed. A further indication as to the flexibility of this contract is that it may be used on contracts based on either a target cost or a lump sum.

As previously mentioned, the CE is designed to be used as the standard form of contract between numerous organisations involved with a project. To facilitate this flexible approach, the two parties to the contract are referred to as the 'Purchaser' and the 'Supplier'. When the client uses this contract to engage professional advisers, or a contractor, he will be identified as the Purchaser, while the advisers and contractor will be identified as suppliers. However, when the contractor uses this form to sub-let portions of the work or to purchase materials and plant, he will now become the Purchaser, and the sub-contractor or supplier will be identified as the Supplier. Where a Supplier is engaged only as a professional consultant (i.e. an architect providing design advice or a QS providing cost advice) they are not subject to the entire contract conditions (see Part 1.1). It is easy for a consultant to identify the clauses that do not apply as they are clearly marked in the contract conditions through the use of shaded text boxes.

There are no specific contract documents detailing the service to be provided by the Supplier, therefore the necessary information is to be provided within the Contract Particulars. In Part 2, there is a section that identifies the Purchaser's brief by means of listing all the relevant documents and their date of issue. Copies of all these documents must be attached to the contract. In return, the service to be provided by the Supplier may be described in Part 3 and/or reference may be made to specific documents and their date of issue. Where documents are identified, copies must be attached to the contract.

Format

As with all current JCT contracts, the Conditions of the CE are broken down into a number of separate sections but, as can been seen from the list below, some of these sections are quite different from the traditional JCT contracts (e.g. Sections 2 to 4):

> Articles of Agreement
> Conditions:
>> Section 1: Definitions and Interpretations
>> Section 2: Working Together
>> Section 3: Primary Obligations of the Purchase
>> Section 4: Primary Obligations of the Supplier
>> Section 5: Allocation of Risks
>> Section 6: Measurement of Performance
>> Section 7: Payment
>> Section 8: Insurance
>> Section 9: Assignment, etc.
>> Section 10: Termination
>> Section 11: Dispute Resolution
>> Section 12: Supplementary Conditions.

Although the CE comprises 12 sections of conditions, these make up only about half the contract document; the other half comprises the Articles of Agreement, which contain the Contract Particulars. It is in the Contract Particulars that some of the unique aspects of the CE may be found.

Contract Particulars

In the CE, the Contract Particulars contain a large amount of information relating to the project, and its content is considerably different from the majority of the JCT standard forms of contract.

Part 1: Details Relevant to the Conditions

This section provides some of the detail more commonly found in other JCT contracts, such as names of the parties, dates for possession and completion, liquidated damages, insurance requirements and the Rectification Period. However, further information is provided on items that are not always evident in other JCT contracts. Some of these items are then subsequently expanded in Parts 2 to 9 of the Contract Particulars.

Part 2: Brief

This identifies the documents that comprise the client's brief. A table is provided where the title of each document is inserted, along with its date of issue. All identified documents are to be attached to the contract.

Part 3: Services

A space is provided to describe the service to be supplied by the Supplier. Again, any documents used to describe the service are to be identified and attached to the contract.

Part 4: Supplier's Key Personnel

In Table A, details of the Supplier's key personnel are to be inserted. Table B may be used to provide details of the supply chain key personnel, if required. Table C provides the names and details of the Project Team.

Part 5: Risk Allocation Schedule

This provides two risk allocation proformas, i.e. Risk Allocation Schedule A and Risk Allocation Schedule B. Either schedule may be used, depending on which format the parties choose.

Part 6: Key Performance Indicators

This identifies the key performance indicators that may be used in this agreement in order to assess the performance of the Supplier and/or the Purchaser.

Part 7: Payment Terms

This identifies whether the agreement is based on a target cost or contract sum basis. If the target cost option is to be used, a considerable amount of detail is then to be provided to demonstrate how the payment terms are to be operated.

Part 8: Third Party Rights and Collateral Warranties

This sets out the third party rights to be provided by the Supplier and the method by which they are to be provided. It also identifies any third party rights that are to be provided by a sub-supplier of the Supplier.

Part 9: Supplementary Conditions

This is a blank section where it is possible to attach any supplementary conditions that are to be applied to the agreement.

Conditions

Section 1: Definitions and Interpretation

Because of the innovative approach adopted by the CE, this section can be very useful as it provides brief explanations of words and phrases not normally found in the more traditional JCT contracts, e.g. actual cost, excluded cost, guaranteed maximum cost, key performance indicators, project team, risk allocation schedule, supplier's margin, supply chain, target cost, etc. Another innovation is that the contract recognises that the parties may wish to use email as a means of giving notices (clause 1.5). However, the contract does not provide a protocol for the use of email and the parties will have to prepare their own protocol for inclusion in the contract, e.g. through the use of a supplementary condition in Part 9 of the Contract Particulars.

Section 2: Working Together

This section commences with the important statement that the overriding principle of this agreement is that the Purchaser and Supplier will work in a collaborative manner, and that this principle is to be extended to other project members. The Project Team is identified in this section and will initially comprise the Client, Lead Designer, Lead Supplier and other Suppliers identified by the client who have a key role to play in the project. The composition of the Project Team may be altered during the progress of the project, with the agreement of the Project Team members. The role of the Project Team is to take on the responsibility for the design and construction of the project and to work towards a successful delivery of the project. To encourage a collaborative approach to the project, it is suggested that the Project Team may consider drawing up a project protocol, setting out the aims and objectives of the Project Team with reference to how working relationships should be developed between the members, and how the project is to be delivered. If a protocol is produced, it is to be prominently displayed in all work areas. However, any such protocol is merely a statement of intent of how the parties intend to co-operate on the project and reinforces the principles of partnering. The protocol is not contractually binding and a party cannot be in breach of contract if they fail to adhere to the terms of the protocol. Nevertheless despite the non-contractual status of the protocol, it is important to note that, in any dispute that is referred to the courts or adjudication, etc., the failure of any party to adhere to the 'Overriding Principle' (i.e. to work collaboratively) should be taken into account when making any award (clause 2.9).

Section 3: Primary Obligation of the Purchaser

As previously mentioned, the Purchaser may be the client, the contractor or a sub-contractor, depending upon where, within the supply chain, the CE is being used. Whoever the Purchase may be, it is their obligation to provide the Supplier with information to allow them to carry out their work and respond to requests for information from them. The Purchaser is to appoint a Purchaser's Representative, who has authority to act on behalf of the Purchaser and is to be the Supplier's first point of contact with the Purchaser. On completion of their works, the Supplier (where they are not a consultant) may request a certificate confirming the Completion Date. If the supply works pass inspection, and any necessary commissioning work or tests, the Purchaser must issue the requested certificate.

Section 4: Primary Obligations of the Supplier

Depending upon the circumstances, the Supplier may be an architect, QS, contractor or sub-contractor. The Supplier is required to provide the Purchaser, or other members of the Project Team, with information regarding the service they are providing. Where a Supplier is providing a professional service, there is a choice in the level of liability the Supplier must accept with regard to their 'duty of care'. There is a provision for the Supplier to 'exercise reasonable skill and care in carrying out the design and/or other professional services for which he is responsible'. Or, alternatively, the Supplier may be required to accept the liability to 'exercise the level of skill and care reasonably to be expected of an appropriately qualified and competent professional designer providing those services in relation to projects of a similar size and scope to the Project'. Where a project is large and complex, the use of the second alternative would impose a greater obligation on a Supplier than the first alternative, since the Supplier would be deemed to have a greater knowledge and competency, in relation to the project than a normally qualified and competent professional.

The Supplier is responsible to the Purchaser for the delivery of his Service as set out in the contract and he is also responsible for the performance of his Supply Chain. Furthermore, a Supplier may be identified as being the 'principal contractor' in accordance with the CDM Regulations. A Supplier may also be identified as being responsible for preparing and maintaining the project programme and/or being responsible for arranging progress meetings. To ensure that the project proceeds on a collaborative basis, a Supplier is expected to appoint his Supply Chain using the CE or an equivalent form of contract. The Supplier is required to comply with all reasonable written instructions from the Purchaser. A Supplier is also expected to take part in 'project planning, risk and value engineering workshops involving all or relevant members of his Supply Chain and other Project Participants as necessary'.

The standard of the Service supplied by the Supplier is to be to the Purchaser's reasonable satisfaction. The Supplier (i.e. personally or through his Supply Chain) is expected to provide the personnel, materials and plant necessary for the completion of the Service. Goods and materials are to comply with the contract requirements. The Supplier's key personnel, and the key members of the Supplier's supply chain, are to be identified and named in tables within the Contract Particulars. The Supplier is required to ensure that these people and organisations are engaged in the delivery of the Service. However, a Supplier may change his key personnel and key members, with the Purchaser's approval.

Any Supplier who provides a service that is more than just a professional consultancy service is subject to a Rectification Period. The default Rectification Period runs for 12 months from the completion date, although a different time period may be inserted in the Contract Particulars.

On occasions, a Supplier may be designated as a Lead Designer or a Lead Supplier. A Lead Designer is responsible for co-ordinating the design input of all the other Suppliers, but is not liable for the design of the other Suppliers. Similarly, a Lead Supplier is responsible for co-ordinating and integrating the work of the other Suppliers.

Section 5: Allocation of Risks

Another example of how the CE endeavours to encourage a collaborative philosophy amongst the Project Team is the incorporation of a risk register into the contract. A Supplier may be identified as the party responsible for the preparation of the risk register and, as a consequence, would be responsible for ensuring that an initial risk assessment is carried out. The other Suppliers are expected to co-operate with the risk assessment process. A Supplier may also be identified as the party responsible for keeping the risk register up to date and, as such, will be responsible for calling regular meetings with the Client and Project Participants to review and update the risk register. The purpose of the risk register is to identify events that may arise during the project and which would have an adverse effect on the project. These risks would be analysed to assess the probability of their occurrence and the potential impact on the project. From this information it is possible to prepare an action plan in response to the identified potential risks.

There is the option for the production of a risk allocation schedule. If a risk allocation schedule is to be used, the CE provides a choice of format. A schedule may be produced where each individual risk is assessed for time and cost consequences (Schedule A). Alternatively, each risk may be individually identified, but the cost and time consequences are each added together to produce just one total figure for time and another for cost (Schedule B). The risk allocation schedule is an important document if the Purchaser becomes responsible for making a contribution towards the cost of identified risks, or for agreeing an extension of time (see clause 5.3.2). The notes in the Guide[7] recommend that a risk allocation schedule should be used in contracts between the client and the contractor who is providing the majority of the construction services. The contractor is likely to carry out a similar exercise with his major sub-contractors but may not consider it a realistic exercise with the smaller sub-contractors. The purpose of the risk allocation schedule is to identify what proportion of the risk (measured in cost and time) is allocated to the Purchaser and/or the Supplier. The schedule is a contract document and should not be changed during the progress of the project unless there is agreement in writing from both parties to make an amendment.

The CE makes provision for a Supplier to claim additional time and/or costs in response to a 'relief event'. A relief event in the CE tends to give a Supplier similar rights to an extension of time and payment for loss and/or expense or variations as are found in the SBC/Q. There are four relief events identified:

1. Instructions from the Purchaser to change the Service or Project
2. An act or omission of the Purchaser, etc.
3. The occurrence of a risk identified in the risk allocation schedule
4. The occurrence of a risk not identified in the risk allocation schedule, which was not foreseeable at the start of the contract and is beyond the control of the Supplier

In connection with item 3 above, a Supplier's right to claim additional monies or time may be negated or limited depending upon the allowances already included in the risk allocation schedule and whether or not the Supplier has been identified as the party responsible for this risk. With reference to item 4, a Supplier is entitled to recover only a percentage of the time and cost associated with the risk. The actual percentage should be inserted into the Contract Particulars. The default position is for the time and cost to be shared equally between Purchaser and Supplier.

There is an obligation on both Purchaser and Supplier to inform each other as soon as either becomes aware of a relief event. Once a relief event is notified, it becomes the responsibility of the Supplier to prepare a statement detailing the impact the event will have in terms of time and cost. In accordance with the collaborative philosophy of the CE, the Purchaser and Supplier are to work together to try and minimise the effect of the relief event and to reach agreement on the financial and time consequences. If it is not possible to forecast the effect a relief event may have on a project, the Supplier and Purchaser may agree an estimate based on assumptions. If the assumptions are subsequently found to be incorrect, the parties may agree to correct the original estimate.

Section 6: Measurement of Performance

It is the intention of the CE that the Purchaser and Supplier should monitor each other's performance against a set of identified 'Key Performance Indicators'. The parties should carry out formal reviews on a regular basis to measure current performance and discuss means by which each other's performance may be improved. In accordance with a Project Team Agreement (clause 2.2), it is possible for the performance of the Purchaser and Supplier also to be monitored by the Project Team. The Key Performance Indicators to be used in the monitoring process are to be inserted into Part 6 of the Contract Particulars. It should be noted that it is possible to remove the performance review procedure from the contract through an appropriate deletion in the Contract Particulars. The JCT Guide[8] suggests this might be considered where the cost of administering the process is disproportionate to the contract value, e.g. when a contractor has engaged a sub-contractor for a small and minor part of the works.

Section 7: Payment

There are two methods by which the cost of a Service may be determined, i.e.:

* The Target Cost Option
* The Contract Sum Option

The chosen method must be stated in the Contract Particulars.

Where a target cost is to be used, there is also the option to have a 'Guaranteed Maximum Cost'. Under the target cost option, the Supplier is normally entitled to receive payment for his actual costs on a monthly basis, and on top of that will receive payment for the Supplier's Margins at the times identified in the Contract Particulars. If the actual cost of the service comes above or below the target cost, the Purchaser and Supplier will share the overspend or saving. The manner in which the overspend or saving is to be proportioned between the Purchaser and Supplier is set out in the Contract Particulars. The parties may agree just a very simple 50/50 split or it is possible to have different percentage splits for the overspend and savings, and different percentages depending on the value of the overspend or saving. Where in addition a guaranteed maximum price has been agreed, the Supplier will be responsible for any actual costs that are in excess of the guaranteed maximum price.

Example:

Target price is set at £30 million with an agreement to share any saving or overspend 50/50 between Supplier and Purchaser. There is a guaranteed maximum price of £35 million. The actual costs for the project are £36 million. The Supplier will receive £30 million for the target price, £2.5 million for the overspend up to the guaranteed maximum price, and will have to fund the £1 million in excess of the guaranteed minimum price himself.

Where the contract sum method is used, there is an option for the whole service to be remeasured as the work proceeds, or upon completion. The timing of payment is detailed within a schedule completed in the Contract Particulars, i.e. on set dates, upon the completion of activities, or when milestones have been achieved. The Supplier is to submit an application for payment along with a statement confirming the progress made to date.

Whichever of the two payment options apply, the Purchaser, in accordance with the HGCRA, is required to provide proper notices stating the amounts that are to be paid to the Supplier, and likewise if monies are to be withheld. A Purchaser is also liable to pay interest on late payments.

Suppliers who provide more than just a professional service may be liable for liquidated damages, although there is an option to delete this condition. Where liquidated damages are to apply, the appropriate rate(s) should be inserted into the Contract Particulars. To avoid any legal confusion, it is stated that, if the liquidated figures are not inserted into the Contract Particulars, then unliquidated damages are to apply (see Temloc v Erril, 1987). Another innovation for this contract is the option to pay a Supplier a bonus for early completion.

Section 8: Insurance

A Supplier has a similar liability under the CE as he would have under the SBC/Q. A Supplier is liable to indemnify the Purchaser for injury to or death of any person caused by the carrying out of the service, unless caused by the Purchaser's negligence. A Supplier is also liable to indemnify the Purchaser for damage to property (generally excluding the project works and site materials) where caused by the Supplier's negligence. It is possible to place a financial ceiling on the Supplier's liability through Part 1 of the Contract Particulars. However, a Supplier's liability for death or personal injury is specifically excluded from this limitation of liability.

The insurances that are required to be taken out by the Supplier and Purchaser are identified in the Contract Particulars. For a Purchaser, the only insurance identified in the Contract Particulars is a requirement for Property Insurance. A Supplier may be required to take out the following:

- Professional Indemnity Insurance
- Public Liability Insurance
- Contractor's All Risk Insurance
- Employer's Liability Insurance

The CE provides only a very broad outline as to the types of insurance cover that may be provided; therefore, when it comes down to the finer detail of the policy requirements, this information must be provided in separate documents at the tender stage. Another limitation of the CE insurance provisions is the way in which it deals with sub-contractors, i.e. parties who will be engaged as Suppliers by the contractor, who, in this instance, will be acting as the Purchaser. Will all the sub-contractors be required to take out their own Contractor's All-Risk insurance or will they be covered by the insurance taken out by the contractor in his contract with the client? Although the CE insurance provisions may appear to be far simpler than those contained within SBC/Q, the problem is that all the detail will have to be dealt with through other documentation, which could result in a lot of non-standard requirements and more time being required to check through all the documentation. One possible solution may be through the use of a Project Insurance policy, which is an option within the Contract Particulars. Such a policy would integrate all the insurance requirements of the project and provide cover for all parties involved on the project. These policies are not commonly used at present, but this may change in the future.

Section 9: Assignment, Subletting, Third Party Rights and Collateral Warranties

This section is very brief compared to the SBC/Q and contains very little in the way of detail within the conditions, although a little more detail may be found in the Contract Particulars. There is a simple statement to the effect that the Supplier shall comply with Part 8 of the Contract Particulars and ensure that the relevant third party rights and/or collateral warranties are provided, including third party rights or collateral warranties from sub-suppliers providing a design service. There are some slight differences from the SBC/Q; e.g. a Supplier is not allowed to sub-let the whole of the service. Another example is that a Purchaser (when he is also the client) may, without the Supplier's consent, assign both the benefit and burden of the contract to another organisation, who will maintain the functions of the client. If such an assignment takes place, the Supplier, if requested by the Purchaser, is to enter into a 'deed of novation' with the assignee.

Section 10: Termination

The CE and SBC/Q use different wording with regard to termination; in the CE, a party's contract is terminated, while in the SBC/Q (and other JCT contracts) it is the party's employment that is terminated. The JCT is of the opinion that the different wording may have legal consequences.[9] They advise that, where a party's employment is terminated, the contract conditions will remain in force to allow the consequences of the termination to be dealt with between the parties. Where a contract is terminated, the JCT advises that a claim for damages for a breach of contract may be the only remedy for the injured party.

A Purchaser may terminate a Supplier's contract where a Supplier has failed to remedy a breach of contract or when a Supplier has become insolvent. Additionally, where the Purchaser is not the client and their contract is terminated, they are entitled to terminate their Suppliers' contracts. This reflects a similar term in the SBC/Q, namely that, when a contractor's employment is terminated, a sub-contractor's employment is automatically terminated. A Supplier may terminate the contract where the Purchaser has failed to make proper payment or is in breach of contract or has become insolvent.

Following a termination of contract, the Supplier will be paid under one of the two alternative payment procedures. Where the contract is terminated because of the Supplier's default, the Purchaser may withhold any payments due until the cost of completing the service has been calculated. Where the service is completed at an extra cost, the Purchaser may charge those costs against any monies withheld from the Supplier. However, where the contract is terminated by the Supplier, the Purchaser must promptly pay any monies due to the Supplier up to the date of termination. If the Purchaser requests the Supplier to carry out any works after the termination, then prompt payment is to be made for these services.

Section 11: Dispute Resolution

The dispute resolution procedures are generally similar to the SBC/Q but with a number of innovations. Under the CE, both parties are required to notify each other of any anticipated disputes. Once again this is a reflection of the collaborative approach of the CE and is an attempt to encourage the parties to try and resolve problems at the earliest opportunity. Where a dispute does arise the parties are encouraged to try and resolve it through negotiation between their 'senior executives' who are named or identified in the Contract Particulars. If these negotiations reach an impasse, the parties are encouraged to consider the use of mediation. If either the Purchaser or Supplier is a member of the project team, they must inform the team of any dispute that has not been swiftly resolved and keep the project team informed. Either party may refer a dispute to adjudication in accordance with the provisions of The Scheme for Construction Contracts (England and Wales) Regulations 1998.

Section 12: Supplementary Conditions

This is a blank section within the CE, but it allows the parties to incorporate any additional conditions they may wish to apply to the agreement.

Project Team Agreement (CE/P)

The JCT has produced a project team agreement to complement the CE documentation. The purpose of the Project Team Agreement (CE/P) is to reinforce the collaborative principle of the CE by bringing the members of the project team together under a legal agreement. However, the parties' liability under this agreement is limited to the 'Risk and Reward Sharing Arrangement' in Section 3 of the CE/P. The parties to the CE/P will be the client and the Suppliers who comprise the project team. The names and particulars of all these parties are to be found at the front of the CE/P. There is space within the CE/P to insert the names and details of up to five Suppliers, although this number could be increased by attaching further particulars to the agreement. The project team may comprise the client and his key professional adviser(s), his key supplier (e.g. main contractor) and a number of sub-suppliers (e.g. sub-contractors and suppliers) who are responsible for major elements of the project.

The CE/P is not to be used as an agreement on its own; it is to be used as a supplement to the contract entered into by the various members of the project, i.e. the CE. To confirm this fact, it is stated that the CE/P does not supersede the conditions of the CE, therefore the parties that enter into this agreement will be bound by both the CE and the CE/P. It is possible that the CE/P may impose additional rights and obligations on the project team, and should these conflict with rights and obligations contained in the CE, the CE/P will take precedence (see CE clause 2.2).

One of the important aspects of the CE/P is to formalise the collaborative work of the project team, and this is achieved through Section 2 where the parties reaffirm their commitment to the 'Overriding Principle' contained within the CE. Section 2 then proceeds to flesh out the administrative role of the project team and enlarges on the detail contained in the CE. Information and procedures are set down regarding the role and function of the project team, the organisation of meetings, and identifying the individuals who are to represent the project team members.

Section 2 contains an important condition limiting the liabilities of the parties who enter into the agreement (CE/P clause 2.9). The parties who enter into the CE/P do not owe each other a duty of care, either in contract or tort. Similarly, the parties have no liability to each other under the CE/P for any acts or omissions. However, this exclusion of liability does not extend to Section 3, which is an optional section of CE/P and deals with a 'risk and reward sharing scheme'. Failure to comply with this section could lead to legal action.

If a project has been let as a target cost contract, the client may wish to bring the risk and reward procedures together under the umbrella of the CE/P. If the client completes Part 4 of the CE/P, then the optional Section 3 'Risk and Reward Sharing Arrangements between Members of the Project Team' will become operational. The information to be provided in Part 4 is: the project target cost, comprising an estimate of the construction costs and professional fees, the proportion of any surplus (benefit) or deficit (liability) that is to be shared amongst the Suppliers of the project team, and the maximum benefit or liability of each Supplier. Section 3 of the CE/P provides the detail of when the project target cost should be agreed and how it may be adjusted, and then how the project final cost is to be calculated and the date for it to be agreed. If the target cost is exceeded, each Supplier is liable to pay the client their share (up to the maximum liability shown in Part 4); if there is a saving on the target cost, each Supplier is entitled to receive, from the client, a share of that saving up to the maximum benefit shown in Part 4. Any payment, to client or Supplier, is to be made within 20 business days from the date of calculating and notifying each party of the saving or overspend relevant to their service.

Summary

The CE form is based upon a previous non-JCT standard form (i.e. the Be Collaborative Contract), which explains why its style and content differs from the traditional JCT documentation. It is a versatile contract in that it may be used in agreements for the execution of construction work, the supply of professional services or the supply of plant and materials. Within a project it is expected that all the major parties and the supply chain will be engaged under an identical CE form of contract.

The CE form is recommended where the client wishes to use a partnering approach, and as a consequence many of the contract conditions have been drafted to encourage the parties to adopt a collaborative approach to the project. As a reflection of the partnering nature of the CE, there is a procedure for the provision of a risk register with the option of producing a risk allocation schedule.

The works may be let on a contract sum basis, but with an option for the works to be remeasured. Alternatively, the work may be let on a target cost basis with the option to have a guaranteed maximum price.

Prime Cost Building Contract (PCC)

The basic principle of a prime cost building contract is that a contractor will receive payment for his construction costs plus an additional payment for his profit and over-heads. From a contractor's viewpoint this is a very low-risk form of contract as he is virtually guaranteed to recover all his construction costs and receive an agreed fee to cover his overhead costs and profit. As a result, one of the major criticisms of this type of contract arrangement is that there is little incentive for a contractor to be efficient, with the result that a client may pay over the market price for his construction project. It is because of this lack of financial control and certainty that many clients are reluctant to enter into a prime cost contract, and recent surveys undertaken for the RICS show that the contract is rarely used. For example, from the questionnaires returned for the 2004 survey,[10] there was evidence of this form of contract being used on only three occasions. This was for works valued between £100,000 and £500,000 and accounted for just 0.02% (by value) of the construction work captured by the survey. However, despite the criticisms of this form of contract, there are occasions where its use may be advantageous to a client. There have been instances where commercial and retail premises have been damaged through fire, flood or terrorist attack and the owners of these properties have used a prime cost contract to enable the repairs to be carried out as quickly as possible. In these examples, the clients' main priority was to have the works carried out swiftly; the potential loss of revenue from their commercial activities would far outweigh the slightly higher costs they might incur under the prime cost contract.

The advice provided by the JCT is that clients may consider using the PCC where they wish to make a swift start on a project and where there is not enough time to detail the work to allow the use of a lump sum contract, or even a remeasurement contract. Where an Employer uses the PCC, he is required to provide details of the work to be carried out, i.e. the nature and location of the work, a specification, and drawings, although the last item is optional. To assist with the administration of the works, the Employer is required to appoint an A/CA and a QS.

Contract Particulars

The Employer is required to provide an estimate of the prime cost of the work to be carried out and this figure is to be inserted in the Contract Particulars, along with details of how the estimate was built up (see Fourth Recital). It is possible for the Employer to have works carried out by other contractors concurrent with the main contract works. These works are referred to as 'excluded works' and are identified in the Third Recital. It is also possible for further documents to be provided to clarify the scope of the main contract works. Under a prime cost contract, it is frequently difficult to fully define the extent of the works at the contract stage. However, it is important to provide as much information as is practicable so as to reduce the possibility of disputes arising later to the effect that the scope of the works has been exceeded and as a result the contract fee should be increased. To assist in the financial management of the project, the Contractor is to provide details of his onsite staff and his directly employed workforce.

Within the Contract Particulars it is possible to provide lump sum prices or rates and prices to be used in calculating the cost of certain identified items of work. In such instances, the work will be valued in accordance with the lump sum or rates provided, instead of on a prime cost basis.

Carrying Out the Works

As usual, the Contractor's obligation is to carry out and complete the work in accordance with the Contract Documents and in a workmanlike manner, although the Contractor should not commence any work until instructed by the A/CA (see clause 3.14). Because of the nature of a prime cost contract, the Contractor is also required to carry out the works as economically as possible and not to employ more persons on site than is reasonable. These last two requirements are obviously included in response to the criticism that prime cost contracts do not necessarily encourage a contractor to work efficiently.

As under the SBC/Q, the Contractor may sub-let all or part of the works with the written consent of the A/CA, although there are two further conditions with which the Contractor will normally have to comply (clause 3.9). He must obtain a written notice from the A/CA (unless the A/CA has instructed otherwise) approving both the terms of the invitation to tender to be sent out to any sub-contractor and the basis by which the sub-contractor is to be paid.

As under the SBC/Q the Contractor is to comply promptly with any valid instruction issued by the A/CA. However, an exception to this is where the Contractor considers an instruction will result in an alteration to the scope of the works. In this instance, the Contractor need not comply with the instruction as long as he writes to the Employer asking for the Contract Fee to be adjusted to compensate for the change in the scope of the work. It is necessary to refer to Schedule 2 of the contract to see how the Contractor's written application is dealt with. First of all, the A/CA must review the application to determine whether it would be 'fair and reasonable' to adjust the contract fee. If the A/CA considers the contract fee should be adjusted, and if the Employer writes to the Contractor confirming his agreement to the adjustment, then the A/CA (or the QS under instruction from the A/CA) shall, after consulting the Contractor, make an appropriate adjustment to the contract fee and set the date from when the revised fee will be applicable.

Where a revised fee has been agreed, or where the A/CA considers that a revised fee is not applicable, the Contractor must comply with the instruction in question. Where the A/CA considers a revised fee should apply, but the Employer does not agree to the revision, then the Contractor need not comply with the instruction.

There is a provision that allows the A/CA to issue variation instructions in the same manner as under the SBC/Q, but there are no subsequent rules detailing how a QS may be required to value the work. Such rules are obviously superfluous in the PCC, as the cost of complying with the variations will simply be added to the Contractor's Prime Cost.

As in most JCT contracts, there is a provision for the Employer to engage and pay external contractors to carry out work where the original contractor has failed to comply with an instruction of the A/CA, and for the Employer to recover any additional costs from the defaulting contractor (clause 3.11). However, because of the nature of the PCC, it is difficult for the Employer to calculate what extra costs he may have incurred through this process as there are no contractual rates to help determine what price the original contractor would have charged for the work. The QS, therefore, is given the task of assessing the cost to the Employer of having the work carried out by external contractors and of estimating what the cost would have been if the original contractor had carried out the work. If the cost of carrying out the works is higher than the estimate of the QS, then the difference is to be deducted from the Prime Cost.

The A/CA may issue an instruction to exclude any Contractor's personnel from the site, but the PCC adds a warning that the A/CA may not use this instruction to remove Contractor's personnel from site because he considers they are working uneconomically or because they are not required on site. If the A/CA is concerned that the Contractor is not working economically or is keeping too many people on site then he should deal with it through the payment process as explained below.

Payment

With regard to payment the Contractor is entitled to receive monies under three headings:

- The prime cost of the works
- The contract fee
- Any direct loss and/or expense arising from clause 4.16

To assist in the administration of the project the Contractor must maintain appropriate written records of his prime costs and make these reasonably available to the A/CA or QS. If the Contractor fails to maintain the records, or to make them available for inspection, the A/CA (or the QS, if instructed) is entitled to make a reasonable assessment of the prime cost. If the A/CA feels that the Contractor has not been complying with his obligation to carry out the works as economically as possible, or has been employing more personnel than is reasonably necessary (in contravention of clauses 2.1.2 and 2.1.3), he should give the Contractor a written notice of the non-compliance. Subsequently, the A/CA may write to the Contractor to the effect that certain costs associated with the non-compliance will not be allowed as part of the prime cost.

The prime cost of the works is assessed in accordance with the details provided in Schedule 1 (Definition of Prime Cost) and the information is inserted in the Contract Particulars with reference to clause 4.2. It is important to understand what is contained within the Prime Cost of the works, as this is where the majority of the contract costs will be incurred. The purpose of Schedule 1 is to provide a clear indication to the parties of the items that may be claimed under the heading Prime Cost. A basic list of items is as follows:

- Contractor's site staff
- Contractor's operatives that are directly employed and working either on site or off site, i.e. fabrication shop, etc.
- Materials and goods provided by the Contractor for the works
- Plant and services, i.e. machinery, hand tools, scaffolding, temporary works, site buildings, etc.

- Sundries, i.e. fees, rates, fees for temporary services, some insurance premiums, furniture, stationery, etc.
- Sub-contract work

As well as the Prime Cost definitions provided in Schedule 1, it is possible to provide additional detail in the Contract Particulars (see clause 4.2).

The Contract Fee is a mechanism that enables the Contractor to recover costs he has incurred in addition to the Prime Cost, such as his head office overheads, plus an allowance for the Contractor's profit. The Contract Fee may be agreed as a fixed amount or, as an alternative, it may be quoted as a percentage figure; the Contract Particulars will clearly show which method the parties have agreed to use. Whichever method of Contract Fee is used, it is still possible for it to be revised where there has been an alteration to the scope of the works (see Schedule 2, Part 1). On a project where a fixed fee has been agreed, it is also possible to amend the fee where there is a significant difference between the estimated Prime Cost set out in the Contract Particulars (see the Fourth Recital) and the actual Prime Cost of the work carried out. The default situation is that a difference of plus or minus 10% between the estimated and actual Prime Cost would trigger an adjustment to the Contract Fee, although there is the option in the Contract Particulars to insert a different percentage figure. Where the percentage difference has been exceeded, the Contract Fee will be adjusted in accordance with the formula set out in Schedule 2, Part 2 of the PCC contract.

Example:

A project, using the 10% default percentage, has an estimated Prime Cost of £500,000 and a fixed contract fee of £75,000. Towards the end of the project, an interim valuation is prepared where the Prime Cost has been assessed at £600,000, i.e. a 20% increase on the estimated Prime Cost. As a result, the Contract Fee would be adjusted to £82,500 in accordance with the following formula:

$$75,000 \times \frac{100 + (20 - 10)}{100} = 82,500$$

When the Contract Fee is included in the interim certificate, it is not subject to retention, the reason being that the Contract Fee is calculated as a proportion of the net amount of the Prime Cost (i.e. after retention has been deducted) and therefore the retention allowance has automatically been allowed for.

Example:

Estimated Prime Cost = £750,000 and Contract Fee = £90,000

Gross interim valuation £400,000, less retention of 3%, leaves a net amount of £388,000. The Contract Fee payable at this stage is:

$$388,000 \times \left(\frac{90,000}{750,000} \right) = 46,560$$

Under the PCC, the QS is required to prepare an interim valuation before the issue of each interim certificate and the Contractor is required to provide the QS with expenditure details relating to the Prime Cost. By comparison, the SBC/Q requires interim valuations to be prepared only as and when requested by the A/CA, and the submission of a Contractor's application is an option, which the Contractor may use if he wishes. The reason for these differences is that the PCC documentation contains nothing in the way of work schedules or unit rates, and a realistic assessment of the Prime Cost can be achieved only through the Contractor providing his expenditure details to the QS.

As part of the final account preparation, the QS has to prepare a statement detailing how the Prime Cost has been calculated and, if the QS has disallowed any item that has been submitted by the Contractor, he must identify what has been disallowed and explain why. Also, where a fixed sum Contract Fee has been adjusted (because of a significant change to the estimated Prime Cost) a statement must be prepared to show how the adjustment has been calculated. The A/CA must forward all this information to the Contractor.

Summary

The PCC is not a lump sum contract; the Employer will pay the Contractor on a cost reimbursement basis plus a contract fee. The contract fee may be a fixed fee or it may be calculated as a percentage of the prime cost. The PCC is suitable for Employers who want a quick start on site and where there is limited information concerning the actual works to be carried out. There is no provision within the PCC to allow the Employer to pass on any design work to the Contractor.

Measured Term Contract (MTC)

The basis of a measured term contract is that an Employer and Contractor will enter into an agreement to carry out an unknown amount of construction work, which will be paid for by reference to an agreed priced schedule of works. The contract will normally run for a set period of time, e.g. periods of between 12 and 24 months are not unusual. During this period of time, the Employer may issue the Contractor with orders to carry out works as and when required. As the work is carried out or completed, it will be measured and priced in accordance with an agreed schedule of rates. Contracts such as this are ideal for an organisation that has a substantial property portfolio demanding a regular programme of repairs, maintenance and modernisation. For example, a housing association is likely to receive a continuous list of building-related problems from their tenants, such as broken windows, leaking pipes, missing roof tiles, damaged brickwork, etc. Although it may not be financially viable for the housing association to employ its own maintenance division, it would be equally impractical for the housing association to negotiate and agree separate contracts for every item of maintenance and repair work to be carried out. In such a scenario, a measured term contract can be the ideal solution. This fact is confirmed by the JCT, where it is claimed that the MTC is appropriate for use 'where an employer has a regular flow of maintenance and/or minor works (including improvements) to be carried out by a contractor over a specified contract period'.[11]

The MTC, therefore, is a contract suitable for clients who have a continuous programme of repairs and maintenance work, which they wish to be carried out by a selected contractor, over a set period of time and for a price that will be agreed on the basis of a priced schedule of work. By default, it is expected that the National Schedule of Rates will apply to the MTC.

Carrying Out the Works

The MTC is to be administered by a CA; unlike the majority of JCT contracts, it makes no mention of an architect being employed. The length of time for which the agreement will run is stated in Part 2 of the Contract Particulars and, in a footnote to the contract, it states that the period would normally be a minimum of 12 months. However, because of the uncertainties surrounding a contract such as this, there is a provision for either party to bring the contract to an early close, i.e. a 'break clause'. Either party may give a 'break notice' to bring the agreement to a close. By default, a minimum of 13 weeks' notice is required (this may be amended in the Contract Particulars), but no notice can take effect until at least 6 months from the date of commencement of the contract period. The agreement should therefore be guaranteed to run for at least 6 months, assuming there is no prior termination resulting from a contractual default by either party (i.e. clause 8 Termination). Once a break notice has been issued, the Contractor is not obliged to accept any further orders that would require him to work beyond the expiry date of the notice. He must still however accept orders that can be completed before the expiry date and he must complete orders that were issued prior to the notice – even if they cannot be completed before the expiry date.

Because of the nature of maintenance work, it is not possible, at the outset, to clearly identify what works the Contractor will be obliged to carry out; therefore, the contract identifies the geographical area(s) where the work may be required to be carried out, and this is referred to as the Contract Area (see First Recital). Subsequently, the Contract Particulars provide further detail by providing a list of properties where work may be required and a description of the type of work that may be required.

It is not possible to agree a contract sum for the maintenance work that is to be carried out under this contract but, for the Contractor's information, the Employer does provide an estimate of the total value of work it is thought may be required (see Part 5 of Contract Particulars). However, the employer does not guarantee the estimated total value and, if the final value of the work is substantially different from the estimate, there is to be no adjustment to the Contractor's rates.

The principal obligation of the Contractor is to carry out the work that is specified in an order issued by the CA. The work must be carried out in accordance with the Contract Documents and to the standards set out in the order or, if no standards are specified, in accordance with the standards contained in the schedule of rates. As an aid to the Contractor during the tendering process, he is provided with information identifying a minimum and maximum value, which is to apply to all orders that are issued. Any orders issued by the CA must therefore fall within this financial range (subject to any prior agreement between Contractor and CA to work outside this financial range), and any such order must be capable of being completed within the Contract Period (see clause 2.4). When the CA issues an order, it should normally contain a date for commencement of the work and a date by which the work should be reasonably completed. The exception to this is where a system of order priority coding has been set out in the Contract Particulars. An example of such a system is provided in footnote 7 of the contract, e.g. a Code A order may require work to be commenced within 4 hours, and a Code B order within 2 days, etc. If the CA wishes, he may require the Contractor to provide him, at no cost to the Employer, with a programme detailing how the works for an order is to be carried out.

The details for allowing the Contractor access to a work area are to be provided in an instruction from the CA. Because it is envisaged that a large proportion of the Contractor's work may take place in properties that are occupied, the CA is also given the responsibility for ensuring that the occupiers clear the work area as necessary in order to allow the Contractor to carry out the works. If the CA fails to properly comply with these requirements, the Contractor may claim for the cost of any unproductive time resulting from the non-compliance, by way of dayworks.

The CA may issue a variation to the work being carried out by the Contractor. The variation may be issued as an instruction, or it may come into existence by the CA issuing further drawings, directions or explanations.

Payment

The value of the order and any variation work is valued in accordance with Section 5 of the contract. It is normally expected that the majority of the work would be valued in accordance with the schedule of rates. Where the schedule of rates does not cover the actual work carried out by the Contractor, the rates and prices from the schedule may be used to arrive at a fair rate. If it is not possible to arrive at a fair rate through this process, then the value of the work is to be agreed between the Employer and Contractor and, failing any such agreement, the CA is to consult with the Contractor and make a fair and reasonable assessment of the value of the work. Finally, there may be instances where the CA considers the work, or part of the work, is better valued by a daywork, in which case the daywork will be valued in accordance with the detail provided in the contract documentation. A provision that is peculiar to this contract requires the Contractor to give the CA advance notice of the commencement of work, which the Contractor considers should be valued on a daywork basis.

By default, it is the CA who has the responsibility for measuring and valuing the work executed by the Contractor, although this may be amended by using the options provided by item 9 of the Contract Particulars. The options that are available are:

- The CA shall measure and value all orders.
- The Contractor shall measure and value all orders.
- The CA shall measure and value all orders that are in excess of a stated estimated value and the Contractor will measure and value the remaining orders, i.e. those of a lesser estimated value.

If an order has an estimated value in excess of the figure provided in item 8 of the Contract Particulars (£2,500 is the default figure), or the estimated duration of carrying out the order exceeds 45 days, then the Contractor is entitled to receive progress payments. It is the Contractor's responsibility to submit an application requesting a progress payment on an order, and he must allow a minimum of 1 month before submitting another application for a progress payment on that order. The CA is to issue a certificate for payment within 14 days from receipt of an application from the Contractor. Orders that have an estimated value less than the figure stated in item 8 of the Contract Particulars will be paid on completion.

Once an order has been completed, there are two alternative methods to determine the final value of the order. Where the CA has the responsibility to measure and value an order, he must certify its value within 56 days from the date of completion, and payment is then due within 14 days. If the Contractor has the responsibility for measuring and valuing the order, he must submit his application to the CA within 56 days from completion. The CA is to then issue a certificate for that value within 28 days (assuming he accepts the value of the application) and payment is due within 14 days from the date of the certificate.

There is an option to allow this contract to operate on a fixed-price or fluctuations basis. If the Contract Particulars state that clause 5.6.1 applies (this is the default position) then the contract is subject to fluctuations. Normally a new edition of the National Schedule of Rates is published on 1st August each year. On a fluctuations contract the work is to be valued by reference to the schedule of rates current at the time the order was issued.

The Contractor is to write and notify the CA when an order has been completed. If not disputed by the CA within 14 days, the Contractor's notified date becomes the Order Completion Date. If the CA does dispute the Contractor's notification, the CA is to give a written notice once he is satisfied, or has reached agreement, that the order has been completed – and that notice will provide the Order Completion Date.

The Employer is not legally bound to give the Contractor all the maintenance work for the properties in the Contract Area. If he wishes, the Employer may have works carried out by other contractors or by his own workforce. Also, if he wishes, the Employer may provide the Contractor with the materials, plant and equipment required for an order, although the Contractor is then entitled to a handling charge, i.e. 5% of the current value of the resources supplied.

Insurance

With regard to insuring the works and associated risks, the MTC generally keeps to the same format as the majority of JCT contracts, although the works insurance is dealt with in a slightly different manner. Because the MTC has been designed for use on repair and maintenance contracts, there is the usual requirement that the Employer is to take out a specified perils policy to cover the existing structure and contents relating to a work order. This insurance is to be a joint names policy, which is the normal JCT wording or, alternatively, the policy may contain a waiver to the effect that the insurers will not use their subrogation rights to make any claim against the Contractor. However, under the MTC, it is the Contractor who is required to take out an all risks policy to cover the works he is carrying out to existing properties. In most JCT contracts where a Contractor is carrying out works within, or to, an existing structure, it is the Employer who is required to take out the all risks insurance for this work.

Summary

The Measured Term Contract, as its title signifies, is a remeasurement contract. As a consequence, the Employer will not know his total financial commitment at the start of the contract, although the parties will have agreed a method by which the work will be valued. To assist the Contractor, the Employer does provide an estimate of the anticipated contract value (see item 5 in the Contract Particulars) but there is no procedure to allow the contract rates to be altered if the estimated value proves to be inaccurate. The Contract Particulars will state the period of time for which the contract is to run but, because of the uncertain nature of the work, there is provision for either party to terminate the agreement at an earlier date, through the use of a 'break clause'.

Major Project Construction Contract (MP)

Documentation

This is one of the newer contract forms produced by the JCT. It was first published in 2003, when it was given the title of the 'Major Project Form'. The 2005 version of the contract is very similar in detail to the original contract. The main difference between the two forms is that the detail and text of the 2005 version has been altered so that the layout of the contract more closely follows the style adopted by the JCT for their suite of 2005 contracts, although some of the 2005 layout conventions have been ignored. For example, the contract conditions are not broken down into discrete numbered sections, and as a result, the conditions comprise 43 separate clauses. Unlike all the other JCT contracts, the Contract Particulars and Attestation pages are to be found towards the back of the document. Within the attestation section, there is provision only for the parties to enter into the contract as a deed; the JCT considered that clients involved with major projects would normally require the extra protection provided by a deed and therefore did not provide the option of executing the contract under hand. There are no Articles or Recitals and some of the information that would normally be found here has been transferred to the Contract Particulars, e.g. description of the project, the contract sum, details of the CDM co-ordinator.

As its title suggests, this is a form of contract that has been designed for use on major projects. However, it is also designed for use by employers who are experienced clients of the construction industry and who have their own contractual terms and procedures, which they commonly use when engaging contractors. As a result, it is claimed that the MP form is simpler and shorter than the JCT Standard Building Contract forms, although this may well be negated once the Employer has incorporated his in-house documentation into the agreement.

The MP allows the Employer a high degree of flexibility with regard to the method(s) by which the work is to be procured, although it is envisaged that the majority of the work will be let on a design and build basis. One of the key documents for the Employer is the 'Requirements'; this is the document where the Employer sets out what he requires from his Contractor with regard to the works, design responsibility and division of liability between the two parties. The document may provide certain design details already prepared by the Employer, with the remainder to be designed by the Contractor. It is entirely up to the Employer as to how much or how little design responsibility is to be placed on the Contractor. Within the Requirements, the Employer may specify that the Contractor must employ a named specialist to carry out part of the design and/or works, i.e. this may be a consultant or a sub-contractor. The Employer may name just one specialist or he may provide the Contractor with a list of names from which a specialist is to be chosen. The Contractor is responsible for the work carried out by a named specialist. Despite the importance of the Requirements, the contract provides little advice as to the contents of the document, which is why the MP form is recommended for use by experienced clients who would have the knowledge, experience and expertise to ensure that the documentation is properly prepared, although the JCT does provide some recommendations as to the Requirement's contents in their guide to the MP.[12] In response to the Requirements, the Contractor produces his Proposals, which should comply with the detail and demands of the Requirements. If any part of the Proposal differs from the Requirements, the Contractor must clearly bring this to the attention of the Employer so that the Employer may consider the suggestion(s) and, if accepted, they may be incorporated into the Requirements. If the Contractor fails to follow this procedure, then, during the course of the project, the Employer may instruct which is to take precedence – the Requirements or the Proposal. This instruction will not be viewed as a Change and there will be no cost attached to it.

Carrying Out the Works

Unlike under the majority of JCT contracts, the Employer gives the Contractor access to but not possession of the site. As a result, the Contractor does not have an exclusive right to the site and must be prepared for other contractors or persons to be working on site at the same time. The Contractor must complete the works by the Completion Date and any failure to comply with this entitles the Employer to deduct liquidated damages. The level of liquidated damages is to be found in the Contract Particulars, where they must be stated as a daily rate, unlike under other JCT contracts where it is open for the Employer to assess liquidated damages on whatever time basis he wishes. The Contractor does have the right to request an extension of time, although the list of reasons for which an extension of time may be granted is slightly shorter than in other JCT contracts. For example, the contractor may not request an extension of time in response to exceptionally adverse weather conditions or a strike, lockout, etc. that affects the works. On receiving a notification of delay, the Employer has 6 weeks (compared to 12 weeks in SBC/Q) to notify the Contractor of the adjustment, if any, that is to be made to the completion date. A problem sometimes arises when a Contractor is being delayed by concurrent events; where one event would give rise to an extension of time and the other event would not. For example, the Contractor may be facing delays caused by a Change (i.e. claimable) and by the slow work of one of his sub-contractors (i.e. not claimable). It is generally accepted that, in this situation, the Employer should grant an extension of time based upon the claimable event; this principle has been expressly approved by the MP, and the Employer is obliged to follow this approach (clause 18.7.3).

After the Employer has notified the Contractor of any adjustment to the completion date, he is still entitled to subsequently review this notification in response to further documentation received from the Contractor or once it has become evident how the works have actually been affected by the notified cause(s). For example, the Employer may initially grant an extension of time based upon the limited information available at that time but, as more detailed information becomes available, the Employer may review and adjust the previously notified extension. However, in this review, the Employer may not reduce a previously notified completion date unless he has the Contractor's agreement (clause 18.8). Another innovation in this contract is that after Practical Completion, the Contractor has 42 days in which to inform the Employer (and provide relevant documentation) of any further adjustments that he considers should be made to the Completion Date. The Employer then has 42 days to review the Contractor's documentation and adjust the completion date, if he considers it to be fair and reasonable.

It is possible for the Completion Date to be brought forward through an acceleration of the construction works. This procedure must be initiated by the Employer and requires him to invite the Contractor to put forward proposals on whether or not it is practicable for the construction period to be reduced, and, if so, at what cost. If the Employer decides to adopt the proposals (or revised proposals), he may issue an instruction for a Change.

Another innovation of the MP form is that the Contractor is encouraged to review the project with the intention of suggesting amendments that would financially benefit the Employer. Examples that are given in the contract include: a reduction in the cost of the project or of the lifecycle costs of the project and/or an early completion of the project. The incentive for the Contractor to comply with this provision is that he will be entitled to an agreed percentage of the financial benefit; the actual percentage is to be stated in the Contract Particulars and the default figure is 50%. The monies should be received after Practical Completion. Another incentive for the Contractor is an option where the Employer can offer a bonus for early completion. In the Contract Particulars, the Employer may insert a daily amount to be paid upon early completion. For example, if the Contractor were to finish 21 days ahead of the completion date, he would be entitled to receive a payment of 21 times the daily bonus rate from the Employer. If the employer fails to fill in this section of the Contract Particulars, then the default figure is nil. Obviously, for the bonus system to operate effectively, it is important that the parties are able to determine when practical completion has actually occurred. In most construction contracts, the state of practical completion is not defined, and it therefore becomes a matter of judgement and opinion. The MP form has tried to improve on this situation by providing a definition for practical completion (Definitions, clause 1). The definition requires that all statutory requirements and consents must have been complied with; the health and safety file and any required maintenance instructions and as-built information must have been provided. Also any items identified in the Requirements as having to be completed before practical completion must have been completed. Once these criteria have been met, and as long as any remaining minor works do not have a negative effect on the use of the project, the practical completion certificate should be issued.

The main obligation of the Contractor is to carry out and complete the project in accordance with the contract. Under this obligation, the Contractor will have to complete the design, specification and selection of materials as identified in the Requirements. The Contractor is not responsible for the contents of the Requirements, or for any inadequacies with the design contained within the Requirements; any problems in these areas are the Employer's responsibility. The Contractor is, however, responsible for additional design work that he undertakes in accordance with the Requirements. The Contractor's liability for the design work is one of 'skill and care to be expected of a professional designer'. However, a footnote to the contract draws attention to an alternative model clause in the Guide,[13] which, if used, would impose a 'fit for purpose' obligation on the Contractor. The Contractor is required to submit his design documents to the Employer for comment before he carries out the work. This design submission procedure is virtually identical to the submission procedures contained in the SBC/Q and DB contracts. Although the Contractor submits his design documents to the Employer for comment, the Contractor remains liable for ensuring the design and work complies with the contract.

The ground conditions found on site can have an obvious impact on the project design and the Contractor must notify the Employer if he discovers problems with the ground conditions that would require a Change to the Requirements. Whether or not the Contractor receives payment for such a Change depends upon which option has been selected in the Contract Particulars. The default situation is that the Contractor will have to bear the cost of any Change, whereas, if clause 14.2 is selected, the Contractor will be reimbursed for the Change, as long as the ground problem could not have been reasonably foreseen, by a competent contractor, at the base date.

As under the SBC/Q, the Employer has the right to test and inspect the works to ensure that they comply with the contract requirements. He may instruct that non-complying works are to be removed, or that they may remain, subject to compensatory payment by the Contractor. If the Employer discovers an item of non-complying work, he may order further tests, at the Contractor's expense, to determine whether or not there may be similar non-complying work elsewhere on the project. Other JCT contracts that contain this clause also include a code of practice to help the parties to determine what is reasonable in regard to the type and extent of tests to be carried out. The MP form does not contain this code of practice, which again reflects the fact that this form is designed for use by experienced clients and contractors.

After practical completion, the project enters into the Rectification Period where the Employer may instruct the Contractor to return to site and put right any defective work. Under the MP form the Rectification Period is set at 12 months; there is no provision within the contract particulars to insert an alternative period of time. If the Contractor has not remedied any defects within a reasonable time from the end of the Rectification Period, the Employer has a choice of remedies. Firstly, he may arrange for the work to be remedied by someone else, as long as he provides the Contractor with an estimate of the costs. Alternatively, he may notify the Contractor that the defects will not be rectified and provide the Contractor with details of the monies that will be recovered as a consequence of the non-rectification.

Appointment of Consultants

As an aid to the administration of the project, the Employer must appoint an Employer's representative, who has all the power and function of the Employer. The Contractor has to comply with written instructions from the Employer, i.e. the Employer's representative. The Employer may engage other professionals to assist him with the project and the Contractor is to co-operate with these persons but they have no authority to act on behalf of the Employer; it is important that the Contractor remembers this fact. It is also possible for the Employer, through the Contract Particulars, to require the Contractor to employ a consultant who had previously been engaged by the employer. For example, the Employer may have appointed a designer or engineer to help prepare the initial project details, and he now wishes the Contractor to take over this appointment and retain the consultant for the remainder of the project. The procedure for this is set out in the contract conditions. Where the Employer wishes the Contractor to employ a pre-appointed consultant, that person must be identified in the Requirements. Subsequently, when the Employer and Contractor enter into the MP contract, they are to promptly execute a Model Form with any pre-appointed consultant identified in the Requirements. The Definitions in clause 1 explain that the Model Form is a 'model form of novation agreement'. What this means is that the contract between the consultant and the Employer is brought to an end and replaced by a new contract between the consultant and the Contractor. Where a Contractor engages a pre-appointed consultant, he is responsible not only for the consultant's services provided after the date of the main contract, but also for consultancy work carried out before this date, i.e. work carried out for the Employer. One exception to this liability is that the Contractor cannot be responsible for the consultant's input to the content or design contained within the Requirements.

Payment

The Contractor is entitled to receive monthly interim payments. The payments are set out in an interim payment advice issued by the Employer. The date when the Employer is to issue the payment advice is set out in the Contract Particulars; the default date is the 28th day of each month. As a precursor to payment, the Contractor is required to submit a detailed application for payment at least 7 days before the date of issue of an interim payment advice, i.e. by the 21st of each month at the latest. In response, the Employer is to issue an interim payment advice detailing the amount that he intends to pay the Contractor. These monies only become due after the Contractor has sent the Employer a VAT invoice for the amount stated in the payment advice. The final date for payment is 14 days from receipt by the Employer of the VAT invoice. When the Employer makes a payment, he must also include any VAT payment required under the current legislation. After Practical Completion, the Employer is no longer obliged to issue an interim payment advice if the sum due is less than the amount stated in the Contract Particulars; the default figure is £10,000 (see clause 28.2).

The Contractor's application for payment and the Employer's payment advice should be calculated in accordance with the Pricing Document and associated contract sum analysis. There are four different ways in which this pricing document may be prepared, the details of which are to be found in Schedule 2:

1. Rule A, interim valuation
    The Contractor is entitled to be paid the value of work executed up to 7 days before the date of issue of the payment advice. The value of work is calculated from the contract sum analysis provided by the Contractor.
2. Rule B, stage payments
    The Contractor is to receive payment for the stages of work that have been completed up to 7 days before the date of issue of the payment advice. Again these stages should be identified in the contract sum analysis.
3. Rule C, progress payments
    The Contractor is paid on the basis of a priced schedule that is to be attached to the contract. The schedule will identify the monthly value of work that will be completed assuming the Contractor maintains the expected progress. If the Employer considers that the work will not be completed by the date stated in the Contract Particulars, he is to notify the Contractor, informing him of the date when

the Employer believes Practical Completion is likely to be achieved. As a result, the Employer, in consultation with the Contractor, is to make reasonable alterations to the schedule so that it reflects the longer time period now required to complete the works.

4. Rule D, other procedure

There is no set procedure under this rule. It is basically a blank provision that allows the Employer to set out whatever payment procedure he wishes to use to pay the Contractor. Where the Employer decides to make use of Rule D, he must ensure that the details of the payment procedures are attached to the contract.

The Employer needs to specify in the Contract Particulars which of the above payment procedures is to apply to the contract; if nothing is specified Rule A will apply by default.

Changes

During the progress of a project, instructions and events may arise that result in a Change (i.e. what would be considered a variation under SBC/Q). As long as a Change has not come about because of the operation of the acceleration or cost improvement procedures (clauses 19 and 25), the consequences are valued in accordance with one of the two procedures set out in clause 26.

The first option is similar to an SBC/Q Schedule 2 Quotation; the Employer can request the Contractor to provide a quote before a Change instruction is issued but, unlike under the SBC/Q, the Contractor cannot refuse to provide a quotation. The Contractor's quotation must identify the value of the Change and detail any requirements regarding a change to the completion date or a claim for loss and/or expense. If the Employer does not accept the quotation, he must prepare his own valuation of the Change, although the Employer is permitted to request the Contractor to provide a revised quotation.

The second option is to be used where a Change has been identified and notified by one of the parties. Within 14 days of a Change being identified, the Contractor must provide the Employer with details of his proposed valuation, including any necessary back-up information. A valuation under this option must take into account any effect the Change may have on other work and any associated loss and/or expense. It is not possible for the Contractor to submit a separate loss and/or expense claim at a later stage (i.e. under clause 27); it has to be dealt with within the Change valuation. However, any alterations to the completion date are not part of the valuation process and would have to be dealt with as a separate claim under clause 15. On receipt of the details, the Employer has 14 days to send the Contractor his valuation of the Change. The Employer's valuation should be based on the information originally provided by the Contractor and should be in sufficient detail to demonstrate where and how it may differ from the Contractor's valuation. There is no procedure for the two parties to reach an agreement on the Employer's valuation; therefore, it is the Employer's valuation that will be used for payment purposes. If a dispute were to arise over the valuation of the Change, it could be referred to adjudication. However, upon reaching practical completion, the Contractor has a period in which he may review any Changes to the project and consider whether there are further valuations that need to be made. If he considers that further valuations are due, he must inform the Employer within 42 days from practical completion. Upon receiving the information, the Employer is required to review the previous valuations of those Changes identified by the Contractor and inform the contractor, within 42 days, of any further valuation that the Employer considers to be appropriate.

Insurance

Insurance requirements are obviously of major importance on large-scale projects; therefore, it is initially surprising to see that the MP form provides a fairly limited section on insurances. However, this was a deliberate move by the JCT, as the MP form was designed with experienced clients in mind, i.e. client organisations who frequently require project specific insurances that may differ from the standard insurance provisions found in the majority of JCT contracts.

The MP form sets out the basic liabilities of both parties. The Contractor is liable to indemnify the Employer for claims that may arise as a result of injuries to persons or property, unless the injury was caused by a default of the Employer. However, if the Employer has taken out insurance that would cover any of these claims, the Contractor is liable only for any amount not covered by the insurance. If the insurance policy contains an excess, this is the Employer's responsibility; the cost of any excess cannot be passed on to the Contractor. The Employer is liable to indemnify the contractor for any claims faced by the Contractor as a result of injuries to persons or property that have been caused by an act or neglect of the Employer. Again, the Employer's liability is reduced where the injury is covered by an insurance policy of the Contractor.

The MP form does not identify and describe the types of insurances that are to be taken out by the parties. Instead, there is provision within the Contract Particulars where the Employer may list the required insurances, the party responsible for providing and maintaining the insurance, and the amount of any policy excess. The actual details of the insurance requirements will be set out in separate documents, which must be attached to the contract. Under clause 34, the Employer does have the option of requiring the Contractor to take out and maintain cover for professional indemnity.

Assignment and Third Party Rights

Some of the procedures relating to assignment are slightly different from those found in other JCT contracts. The Contractor is not allowed to assign the benefit or the burden of the contract without the Employer's consent. This is basically the same as in the SBC/Q, although the wording there refers purely to the assignment of the contract and does not separately identify the burden and benefit of the contract. However, under the MP contract, the Employer is able to assign the benefit of the contract at any time and without having to obtain the Contractor's consent. There is no mention of the Employer being able to assign the burden of the contract in a similar manner, unless it is to a funder. The Employer is able to assign both the benefit and the burden of the contract to a funder at any time and, from the wording of the contract conditions, the Contractor automatically gives his consent to this assignment when he enters into the contract with the Employer.

Third party rights may be provided to purchasers, tenants and funders under the Contracts (Rights of Third Parties) Act and the procedures are similar to those in the SBC/Q. However, there is no procedure for the use of collateral warranties as an alternative to the Act. The reason for this approach is to reduce the amount of paperwork that can be generated by providing collateral warranties to numerous organisations.[14] Surprisingly, there is no provision in the contract to obtain third party rights or collateral warranties from sub-contractors or from the Contractor's consultants. The JCT claim that these procedures are not necessary because the Contractor is wholly responsible for the actions of his sub-contractors and consultants.[15] However, if the Contractor was to go into liquidation, any third party rights against the Contractor would be virtually worthless.

Summary

As its title implies, the Major Project Contruction Contract is for use on major projects but, more importantly, it should be used only by employers who have a good knowledge and experience of the construction process. The contract conditions are light on detail in places as there is the expectation that the Employer will provide the necessary documentation.

The MP is a lump sum contract; the Contractor is to be paid on a regular basis, although the contract conditions provide at least three optional payment procedures from which the Employer may choose. It is the intention of the contract that the works will be predominantly designed by the Contractor, although it is the Employer's decision as to how much pre-contract design work he carries out himself. It is possible for the Employer to engage consultants to carry out the initial work on the project and for the Contractor to subsequently take responsibility for these consultants through a novated agreement. The MP refers to these as pre-appointed consultants. It is also possible for the Employer to name specialists in the tender documents, whom the Contractor is then obliged to employ.

Some innovations in this contract are: a provision for the Contractor to receive a bonus for early completion; a provision for the Contractor to share in any financial benefits achieved as a result of amendments suggested by the Contractor; and, finally, the contract provides a definition of the term 'Practical Completion'.

Conclusion

This section has provided a brief review of a number of JCT contracts. The purpose of the review was to identify the type of construction project where each contract may be suitably used, followed by a brief analysis of the content of each contract to identify the differences and similarities with the SBC/Q. By necessity, a considerable amount of detail has been omitted in the reviews; for a more comprehensive understanding of each contract, reference should be made to the appropriate JCT Guide.

References

1. The Placing and Managements of Contracts for Building and Civil Engineering Work (HMSO, 1964).
2. Contracts in Use, a survey of building contracts in use during 2004 (RICS, 2006).
3. Ibid.
4. Practice Note – Deciding on the appropriate JCT Contract, The Joint Contracts Tribunal Ltd (Sweet & Maxwell, 2007).
5. Practice Note 5 – Deciding on the appropriate JCT Form of Main Contract (The Joint Contracts Tribunal) Ltd (RIBA Companies Ltd, 2001).
6. Contracts in Use, op cit.
7. JCT – Constructing Excellence Contract Guide (Sweet & Maxwell, 2006).
8. Ibid.
9. Design and Build Contract Guide, paragraph 120 (Sweet & Maxwell, 2005).
10. Contracts in Use, op cit.
11. Measured Term Contract Guide, paragraph 2 (Sweet & Maxwell, 2007).
12. Major Project Construction Guide, paragraph 120 (Sweet & Maxwell, 2005).
13. Ibid.
14. Ibid, paragraph 122.
15. Ibid, paragraph 123.

PART 2

An Analysis of the SBC/Q

Introduction

The purpose of this section is to carry out an in-depth analysis of one of the JCT contracts, i.e. the JCT Standard Building Contract with Quantities (SBC/Q). Although the SBC/Q is not so frequently used as it used to be, it has tended to be a basis for many of the other standard forms produced by the JCT. Therefore, a sound understanding of the SBC/Q can be helpful when trying to interpret and operate other standard forms.

The problem that many people have, who are unfamiliar with contract documentation, is a difficulty in understanding the contract conditions. The contract conditions are carefully drafted by legal professionals in a manner to ensure that there is clarity of meaning and lack of ambiguity. Unfortunately, for many lay people the end result is just the reverse as they find some of the long clauses with limited punctuation and numerous cross-references extremely difficult to comprehend. Even seasoned contract professionals sometimes face the same difficulties as evidenced by the number of court cases heard over the years where parties have argued over the exact meaning of a word or phrase contained within a contract.

In this section, the SBC/Q is examined on a clause by clause basis, to explain the purpose and meaning of each clause. Examples are provided to illustrate some of the procedures contained within the contract and finally case law, which has helped to shape the SBC/Q over the years, is identified. Within the SBC/Q, there is to be found a list of definitions and interpretations of certain words and phrases used within the contract conditions (clause 1.1). Where these words or phrases appear in the contract conditions, they are capitalised to identify that their meaning is defined in clause 1.1 of the contract conditions. The same capitalisation convention is used within the following clause by clause analysis.

The JCT Standard Building Contract with Quantities technically comprises two Contract Documents (see definitions, clause 1.1):

- The Articles of Agreement
- The Conditions

Articles of Agreement

This is where basic information relating to the project is to be inserted by the employer or his advisers. Once the first page of the agreement has been completed it is possible to identify the parties to the contract and their addresses, i.e. the Employer and the Contractor.

Recitals

In the following Recitals, the Employer is to provide a brief description of the work to be carried out and identify the location of the work, and for the purposes of this contract this becomes known as 'the Works'. There is also a confirmation that the Employer has had drawings and bills of quantities prepared to explain what work needs to be done. Further administrative information is provided in the Recitals to the effect that the Contractor has supplied the Employer with a fully priced bill of quantities (i.e. the Contract Bills) and that the drawings used for tendering are identified and attached to the contract (i.e. the Contract Drawings). There are also a number of optional Recitals that must be deleted if they are not required. The options relate to whether or not the Contractor is to provide an Activity Schedule, whether or not the Employer is to provide an Information Release Schedule, whether or not the Works are to broken down into Sections and whether or not the Agreement is to contain a Contractor's Designed Portion. Finally within the Recitals, there is mention of the Construction Industry Scheme (CIS). The CIS relates to the collection and payment of income tax within the construction industry, and is usually only of relevance for payments between the Contractor and his sub-contractors. However, in certain circumstances an Employer may be judged to be a contractor in his own right (i.e. the Employer undertakes construction work of his own), which means in this instance the main contractor would have to provide certain tax documentation to the Employer (see Fourth Recital).

Articles

Other personnel involved with the project are identified in the Articles, i.e. the Architect/Contract Administrator (Article 3), the Quantity Surveyor (Article 4) and with reference to the CDM regulations the CDM Co-ordinator (Article 5) and Principal Contractor (Article 6). It is envisaged that normally the A/CA will take on the role of CDM co-ordinator and the Contractor the role of Principal Contractor.

If for any reason the A/CA or QS, named in the Articles, ceases to be employed in the project, then the Employer is to nominate a replacement. The Employer is to renominate a successor within 21 days from when either the A/CA or QS ceased to be employed on the project, but the Contractor does have the right to object to this replacement within 7 days of their nomination. If the Employer does not accept a Contractor's objection, an adjudicator may resolve the dispute in accordance with clause 3.5.

Further key information that is to be found within the Articles is the value of the contract sum (Article 2) and a statement to the effect that the Contractor's obligation is to carry out and complete the Works in accordance with the Contract Documents. The Employer and Contractor may refer any dispute to adjudication under Article 7 (and see clause 9.2) for a quick interim settlement but for a more permanent settlement disputes may eventually be referred to arbitration (Article 8) or litigation (Article 9). A choice has to be made in the contract as to which is the preferred method of settling disputes – arbitration or litigation. The choice is determined by whether or not Article 8 and clauses 9.3 to 9.8 have been deleted in the Contract Particulars, i.e. where these are deleted then litigation is the chosen method. The default position is for litigation to apply.

Contract Particulars

The Contract Particulars is split into two parts. Part 1 of the Contract Particulars is a very important section of the Agreement, it is here that a lot of the detail relating to the administration of the project is to be inserted. As an aid to the administration, the JCT has adopted a policy of including default values into the contract particulars wherever appropriate. For example, if no period of time is specified for the Rectification Period, a period of 6 months will be applied; if no retention percentage is specified, then a figure of 3% will be applied. Part 2 of the contract particulars provides a facility for third party rights and collateral warranties to be arranged in the favour of Purchasers, Tenants and Funders. It is important to ensure that the contract particulars is completed correctly so that the Employer's rights are protected.

(See Bramall & Ogden Ltd v Sheffield City Council, 1983; Temloc Ltd v Erril Properties Ltd, 1987)

Attestation

To confirm their acceptance and agreement to all these details the Employer and Contractor are required to sign the Agreement on the 'Attestation page' at the end of the Articles of Agreement. If the two parties sign the contract on the first attestation page, then they will have 'contracted under hand'; this requires a signature from both the Employer and Contractor or their authorised representatives and for their signature to be witnessed. The legal implication of executing a contract under hand is that either party is liable for any breaches of contract for 6 years from the cause of action, in accordance with the Statute of Limitations Act 1980. If the parties complete the second attestation page, then they will have executed the contract 'as a deed'. In this instance, the names of the Employer and Contractor are inserted in the appropriate place and if the parties have a common seal, they are to use this to make an impression on the page and this is to be confirmed by the signature of two persons authorised to use the company's common seal. Because of changes to company law, it is not now necessary for all companies to possess a common seal, therefore in this instance the fact that a company is executing a contract as a deed is to be confirmed by the signatures of personnel authorised to contract on behalf of the company, e.g. a director and company secretary, or two directors. The legal implication of executing a contract as a deed is that the parties are liable for any breaches for a period of 12 years from the cause of action.

(With reference to the Limitations Act see Applegate v Moss, 1979; Gray v T P Bennett & Son, 1987)

Conditions

Section 1: Definitions and Interpretation

Definitions
1.1   This provides a limited glossary of some of the words and phrases used within the contract. Where these words and phrases occur in the Agreement or the Conditions in a capitalised form, then the definitions provided in clause 1.1 will be deemed to apply. The exceptions to this rule are where the context in which the word or phrase is being used requires an alternative interpretation, or where the contract expressly provides an alternative meaning.

Interpretation
Reference to Clauses etc.
1.2   This provides an explanation of the clause referencing system used within the SBC/Q. The advice provided is that normally where a reference is made to a clause or schedule it is referring to a clause or schedule within the SBC/Q. The exception to this rule is where a reference is made to a clause relating to another document and this fact is clearly stated; for example, in clause 9.4.2 reference is made to Rules 2.6, 2.7 and 2.8; in clause 9.3 it is clearly stated that the Rules are those to be found in the Construction Industry Model Arbitration Rules.

Where within a Schedule (i.e. Schedules 1 to 7 at the back of the SBC/Q) reference is made to a paragraph, it is to be assumed that the reference relates to a paragraph contained within that specific schedule. For example, in Schedule 7 paragraph A.6, reference is made to payments made as a result of 'paragraphs A.1 and A.2 and A.3'. This is a reference to paragraphs A.1 to A.3 in Schedule 7 and should not be confused with the same numbered paragraphs in Schedule 3. The exception to this is where the wording clearly indicates that a paragraph reference relates to a different Schedule.

Agreement etc. To Be Read as a Whole
1.3   The advice provided here is that the various details contained within SBC/Q are not to be considered in isolation. The contract must be read as a whole and the user must be aware and must take account of the fact that some conditions, etc.

may be modified or qualified by other conditions or entries within the contract. For example, the payment rules for testing the Works under clause 3.17 are modified when the A/CA issues an instruction to test work under clause 3.18.4.

It should also be noted that the SBC/Q takes precedence over the Contract Bills and any Contractor Designed Portion Documents (i.e. Employer's Requirements, Contractor's Proposal, CDP Analysis). The Contract Bills and CDP Documents may not be used to override or alter clauses within SBC/Q. However, it is possible to impose additional obligations or requirements on the Contractor through the Contract Bills or CDP Documents as long as they do not conflict with or attempt to alter any clause in the SBC/Q. For example, the Contract Bills may contain a statement to the effect that '*the Contractor is to give the contract administrator 24 hours' notice before covering up any sub-structure works*' – this is valid as it does not contradict any obligation contained within the contract conditions. However, a statement in the Contract Bills such as '*the final date for payment for an Interim Certificate shall be 21 days from date of issue of that certificate*' is invalid as it is in direct conflict with clause 4.13.1.

(See English Industrial Estates Corporation v George Wimpey & Co Ltd, 1972; Moody v Ellis, 1983)

Headings, References to Persons, Legislation etc.

1.4 This section provides basic information on certain conventions used when the contract clauses were drafted, although there is a caveat that the details listed below may be overridden where the context of the wording requires it. The normal conventions are as follows:

1.4.1 The headings in SBC/Q are only provided to help the reader to use the contract; they do not form part of the clauses themselves and as a result cannot have any effect on the interpretation of the clauses relating to that heading.

1.4.2 This is an explanation that where a reference is made in the singular this may also be read in the plural and vice versa. For example, in clause 1.6 reference is made to 'Purchasers, Tenants and/or Funder' – this may also be read as though it were referring to a Purchaser, Tenant and/or Funders.

1.4.3 A clarification that the contract does not relate to any specific gender. In general the contract wording tries to avoid the use of *he* or *his* and refers to persons or uses working titles such as Employer, Contractor, Quantity Surveyor.

1.4.4 Where a reference is made to a 'person' this may be interpreted to include any individual, firm, partnership or corporate body. For example, clause 3.21 refers to an instruction to exclude persons from site, this could mean the exclusion of an individual or may refer to the exclusion of a subcontract company.

1.4.5 Where a reference is made to 'a statute, statutory instrument or other subordinate legislation' it is deemed to be referring to legislation that is current at the time.

Reckoning Period of Days

1.5   Where, under the contract, it is required that an action be carried out within a set number of days from an event or date, then the period of time will start to run from the day after that event or date. For example, clause 4.13.1 states that the final day for payment is 14 days from the date of issue of the Interim Certificate; therefore, if an interim certificate is issued on the 7th of the month, the first day of the payment period would commence on the 8th and the fourteenth day would be the end of the day of the 21st. If one day's public holiday were to fall within this period then the final payment day would be the 22nd.

Contract (Rights of Third Parties) Act 1999

1.6   Under the Contract (Rights of Third Parties) Act, it is possible to provide third parties with the right to enforce terms in a contract to which they are not party. The JCT has allowed for a limited use of the Contract (Rights of Third Parties) Act. It is possible for Purchasers and Tenants, where identified in clause 7A, and Funders, where identified in clause 7B, to be given enforceable rights under this Act. All other third parties are specifically excluded from being able to use the Act to enforce any term within this contract.

Giving or Service of Notices or Other Documents

1.7   On occasions, the contract conditions require that a notice is to be issued (e.g. see clause 2.15) or a document is to be provided (e.g. see clause 2.9.2). If the contract provides no specific instructions on how a notice, etc. is to be given or served, then the following will apply:

1.7.1   the notice or document may be given or delivered by any effective means (e.g. by hand, courier, post, etc.). The notice or document will be deemed to have been properly delivered when given by hand or post to the address stated in section 1.7 of the Contract Particulars. If no information is provided in the Contract Particulars, then the address provided on the first page of the Articles of Agreement is to be used. During the progress of the project, it is possible to agree an alternative address for delivery of notices, e.g. an Employer may move office part way through a project.

1.7.2   If there is no currently agreed address, then the notice may be delivered to a party's last known main business address or, where the party is a limited company, delivery may be made to the company's registered or principal office. Such information should be provided on a company letterhead or obtainable from Companies House.

Electronic Communications

1.8   Although it is necessary to comply with clause 1.7 with regard to certain communications, it is possible for other communications to be made in an alternative manner as set out in the Contract Particulars or, as agreed (in writing), between the parties. If the parties wish to use electronic communications, this must be explained in the Contract Particulars, i.e. which communications may be made electronically and the format to be used. For example, it may be agreed that the construction information and Contractor's master programme detailed in clause 2.9 may be exchanged electronically, in which case it will be necessary to state what software is to be used for the various documents. If section 1.8 of the Contract Particulars is left blank then all communications must be in writing, although the Employer and Contractor can agree to change this during the progress of the project. With the increased use of electronic communications it is important that the parties are fully aware of the requirements set out in clause 1.7. For example, clause 1.7 identifies that certain contract clauses may state a specific means of issuing a notice or document that would preclude electronic communications, e.g. clauses 7.4 and 8.2.3 state that notices must be given in writing and delivered by hand, special or recorded delivery. An email would not be acceptable in this case.

Issue of Architect/Contract Administrator's Certificates

1.9   In most instances any certificates issued by the A/CA will be issued to the Employer with an immediate copy issued to the Contractor.

For advice on what is deemed to be the date of issue of a certificate see Cambs. Construction v Nottingham Consultants (1996).

Effect of Final Certificate

1.10   The Final Certificate, as its name implies, is the last certificate to be issued by the A/CA. The issue of this certificate can be used as conclusive evidence to show that certain aspects of the project have been fully agreed and completed; however, the wording of the clause is not always easy to follow and it can be confusing to understand where the Final Certificate is conclusive and where it has no effect at all.

1.10.1   This part of the clause identifies that the conclusive effects of the Final Certificate will be limited where the parties have entered into formal dispute proceedings (e.g. adjudication, arbitration or litigation). For example, the Final Certificate will not affect any issues that are the subject of adjudication, arbitration or litigation proceedings which commenced before the issue of the certificate (clause 1.10.2) or within 28 days after the issue of the certificate (clause 1.10.3). It is also important to be aware that the Final Certificate will not have an effect on an issue where there is evidence of fraud. Otherwise the effect of the Final Certificate is as follows:

(With reference to fraud, see Applegate v Moss, 1970; Gray v T P Bennett & Son, 1987)

1.10.1.1   conclusive evidence that where, within the Contract Documents, it is expressly described that the quality of goods, materials or standard of workmanship is *to be for the approval of the Architect/Contract Administrator*, then the A/CA is satisfied with the quality of work provided by the Contractor. The reason for this statement is that in this instance the quality standard is entirely a subjective opinion of the A/CA and by issuing the Final Certificate the A/CA is impliedly expressing his approval. It is important to be aware of the severe limitations of this clause. In practice it is rare for the quality of goods and standards of workmanship to be described as having to be *to the Architect's approval*. To confirm the limitations of this clause, the JCT actually states that the certificate has no effect regarding the standards of *other* materials or workmanship; i.e. where goods, materials and workmanship are fully described in the Contract Documents the Contractor will remain liable for these materials, etc. where they do not comply with these objective standards.

The interpretation of this clause caused a degree of confusion in earlier versions of the JCT contracts and following a significant court case in 1994 the JCT reworded the clause to try and clarify the conclusive nature of the Final Certificate (see Crown Estate Commissioners v John Mowlem & Co. Ltd, 1994).

1.10.1.2   conclusive evidence that the Contract Sum has been fully adjusted in accordance with the contract conditions. This clause is qualified to the extent that it is still possible to revise the Contract Sum at a later date if it is subsequently found that work has been accidentally included or excluded or in the light of mathematical errors;

1.10.1.3   conclusive evidence that extensions of time awarded under clause 2.28 have been fully dealt with;

1.10.1.4   conclusive evidence that any claims for payment for direct loss and/or expense under clause 4.23 for matters listed in clause 4.24 have been fully and finally settled. This means, once the Final Certificate is issued and accepted, the Contractor cannot challenge any of the amounts previously paid for loss and/or expense. Also, the Contractor may not submit any further claims for loss and/or expense, even though he may only have realised he was entitled to further claims, after the issue and acceptance of the Final Certificate.

1.10.2   If either party has commenced dispute proceedings before the issue of the Final Certificate then the effects of the certificate are suspended until either:

1.10.2.1   the dispute proceedings are concluded, whereupon the Final Certificate becomes effective but subject to any decisions or awards of the proceedings, or

1.10.2.2   a 12-month period has elapsed since the issue of the Final Certificate during which neither party has taken any further steps in the dispute proceedings. On the expiry of the 12-month period, the Final Certificate will become effective, although it is subject to any agreements and settlements that may have been previously reached in regard to the issues in dispute.

The date when the Final Certificate takes effect will depend upon which of the above two events occurs first.

1.10.3   If either party commences dispute proceedings after the issue of the Final Certificate (and within 28 days of its issue) then the Final Certificate is immediately effective, with the exception of the matters or issues that have been raised in the proceedings.

1.10.4   Where an adjudicator has given a decision after the date of issue of the Final Certificate then either party may still refer the dispute to arbitration or litigation as long as they commence proceedings within 28 days of the date the adjudicator gave his decision.

Effect of Certificates other than Final Certificate

1.11   Apart from the Final Certificate, no other certificate provides conclusive evidence that work, materials or Contractor's design work for CDP is in accordance with the contract. It should be remembered that the conclusivity provided by the Final Certificate regarding the quality of work and materials, etc. is very limited within the very restricted provisions of clause 1.10.1.1.

Applicable Law

1.12   The contract is governed by the law of England. If the parties wish to use another code of law then the existing clause 1.12 will have to be deleted and the appropriate details put in its place.

Section 2: Carrying out the Works

Contractor's Obligations
General Obligations

2.1 This clause sets out the Contractor's overriding obligation to complete the Works in a '*proper and workmanlike manner*' and in accordance with the Contract Documents. The work must also comply with the Construction Phase Plan and other Statutory Requirements. The Contractor is also responsible for making sure that all Statutory Requirement notices are given.

A definition of Statutory Requirements and Works is provided in Definitions, clause 1.1.

See Greater Nottingham Co-operative Society v Cementation and Others (1988) to appreciate why the JCT has incorporated the term '*proper and workmanlike manner*' into the contract.

Contractor's Designed Portion
Contractor's Designed Portion is where the Employer requires the Contractor to undertake the design for a portion of the Works.

2.2 This clause is only relevant where the Contractor is required to undertake a Contractor's Designed Portion (CDP) as identified in Article 7. The clause sets out the Contractor's basic obligations for undertaking the design for the CDP and for its integration into the Works as follows:

2.2.1 The Contractor is to complete the design for the CDP, taking into account any relevant information that may be contained in the Contract Drawings and Contract Bills. Where the standards of materials, goods and workmanship are not specified in the Employer's Requirements or Contractor's Proposals it is the Contractor's responsibility to decide upon the specification and standards to be used.

2.2.2 The Contractor must comply with directions given by the A/CA that relate to the integration of the CDP design with the main Works, for example the A/CA may wish the CDP work to be carried out in a specific manner or sequence to ensure it marries in with the main Works. The Contractor's obligation to comply with such directions is subject to a right of objection through clause 3.10.3.

Note that the A/CA is giving the Contractor *directions* on the integration of the design works, and not *instructions*.

2.2.3 The Contractor is required to comply with CDM Regulations 11, 12 and 18 when carrying out the CDP work.

(*Note*: regulation 11 relates to the duties of designers, and the main purpose of this regulation is to ensure a designer, when carrying out design work, avoids 'foreseeable risks to the health and safety of any person'. Regulation 12 refers to designs prepared or modified outside Great Britain. In this situation the person commissioning the design, if he is based in Great Britain, is responsible for ensuring that regulation 11 is met. If the commissioning person is not based in Great Britain then any client for the project becomes responsible for compliance with regulation 11. Regulation 18 refers to additional duties of designers, and relates to projects that are notifiable (see explanation in clause 3.25). Where a project is notifiable, detailed design work should only be undertaken after the CDM co-ordinator has been appointed. The designer is also required to work with and assist the CDM co-ordinator in his duties.)

Materials, Goods and Workmanship

The purpose of this section is to define the standards of work and materials that the Contractor is to provide for the Works.

2.3.1 This confirms the standards required for the materials for the Works. The standard required is stated in the Contract Bills, i.e. the Preambles Bill. Normally the material standards will be clearly specified by reference to specific manufacturers' products, BS standards or the NBS. Occasionally the standard required may be specified as only being *'to the reasonable satisfaction of the Architect/Contract Administrator'* (see clause 2.3.3).

The standards for materials and goods for CDP work should be described in the Employer's Requirements, but where this is not the case then the standard will be that described in the Contractor's Proposals or Contractor Design Documents issued under clause 2.9.2.

In both of the above situations there is a proviso to the effect that the materials and goods must be actually obtainable. For example, if a manufacturer stops making a specified range of kitchen units the Contractor would not be obliged to scour the country to try and locate any remaining units. For non-CDP work a variation would have to be issued to specify alternative kitchen units. The issue is more complicated with CDP work, i.e. were the kitchen units specified in the Employer's Requirements or was the choice left to the Contractor?

The Contractor may substitute goods and materials, e.g. use plastic rainwater goods from a different company to that specified in the contract – but only with the written consent of the A/CA. The A/CA is required to give a written response within a reasonable time and cannot unreasonably withhold his consent. The JCT provides no advice about a possible price adjustment where a Contractor gains a cost saving by substituting one material for another. Where this situation arises the A/CA could issue a variation instruction confirming the change in specification and a reduction could be agreed when valuing the variation.

(See Leedsford Ltd v The City of Bradford, 1956)

2.3.2 The standard of workmanship is to be as described in the Contract Bills except for CDP works, where the workmanship is to comply with the standard described in the Employer's Requirements or, failing that, in the Contractor's Proposals.

2.3.3   There may be instances where the quality of materials and goods or the standard of workmanship are expressly required to be to the approval of the A/CA, i.e. it is for him to judge what standard is required. In such case it is a requirement that the materials, goods, etc., provided by the Contractor shall be to the A/CA's reasonable satisfaction (see clause 1.10.1.1 to appreciate the implications of this phrase).

Where the Contract Bills fail to describe the standard for goods, materials and/or workmanship, then the obligation on the Contractor is to provide a standard that is appropriate to the type of project that is being undertaken. If a Contractor is engaged on a project for a prestigious hotel with a high-quality specification then he is to provide that standard for all the work where no standard has been specified. Similarly, if the Contractor's Proposals fail to state the standards of some goods, materials or workmanship then the standard provided must be comparable to those standards stated in the Contractor's Proposals. In situations such as this it can be very difficult to reach an agreement on what is an appropriate standard; therefore, it is preferable to ensure that all goods, materials and workmanship are adequately specified at the outset.

2.3.4   If the A/CA wants to check on the quality and authenticity of materials and goods provided for the Works he may ask the Contractor to provide evidence, e.g. delivery tickets, certificates, etc.

2.3.5   The Contractor is expected to take reasonable steps to encourage site operatives to be registered under the Construction Skills Certification Scheme (CSCS) or any similar recognised scheme. There is no express term in the contract to police this requirement. It is seen as an encouragement for the Contractor to engage in good practice and support the use of the CSCS.

Possession

Date of Possession – Progress

2.4   The 'Date of Possession' and the 'Completion Date' are to be found in the Contract Particulars (Sections 2.4 and 1.1, respectively). The Contractor may move onto the site on the Date of Possession and is then required to complete the Works by the Completion Date. He can obviously complete the Works before the Completion Date if he wishes. He is also expected to work in a regular and diligent manner (see clause 8.4.1.2 to appreciate the significance of this term). It is possible for the Works to be broken up into Sections, and each Section may have its own Date of Possession and Completion Date.

(See Glenlion Construction Ltd v Guinness Trust, 1987)

With reference to the Works insurance it is important to be aware of which party has possession of the site during various stages of the project. It is confirmed that it is the Contractor who has possession of the site until the date of issue of the Practical Completion Certificate and the Employer has no right to take possession of any of the Works prior to that date. Similarly, where the Works is broken up into Sections, the Contractor will have possession of each Section from the appropriate Date of Possession until the Section Completion Certificate.

The above rules may be overridden where the Employer uses clause 2.33 to take over part of the Works through partial possession or where the Contractor's employment is terminated under Section 8.

Deferment of Possession

2.5   This is an optional clause that allows the Employer to defer giving the Contractor possession of the site, or Sections of the site. The maximum period of deferment is six weeks from the Date of Possession stated in the Contract Particulars. Six weeks is the maximum time allowed for deferment and it is possible for a lesser period to be inserted in the Contract Particulars. It is important to check the Contract Particulars to see if this optional clause applies or not and, if it does, to see whether a lesser time than 6 weeks has been inserted. There is no information as to how the Employer notifies the Contractor of a deferment but a written notice would be advisable.

(See Rapid Building Group Ltd v Ealing Family Housing Association Ltd, 1984)

Early Use by Employer

2.6.1   It is possible for the Employer to make use of, or to occupy the site or the Works (or any part) before the issue of a Practical Completion Certificate or a Section Completion Certificate. For example, the Employer may wish to store materials, goods or furniture on site, he may wish to operate machinery that has been installed or make use of some of the facilities. This procedure is referred to as *'use or occupation'* by the Employer and should not be confused with partial possession. The Contractor still retains possession of the areas that are being occupied or used by the Employer. The Employer must gain the Contractor's consent to occupy part of the site or Works, but this consent cannot be unreasonably withheld or delayed. The Works insurer must be given prior notification of the Employer's intention to use or occupy the Works and confirmation must be received to the effect that the use or occupation will not affect or invalidate the Works insurance. The insurer will be notified by either the Contractor or Employer, depending upon which one of them took out the Works insurance.

2.6.2   If the Contractor is responsible for obtaining Works insurance under Option A and the insurer requires an additional premium to allow the Employer to use the site, the Contractor is to inform the Employer of the additional premium required. If the Employer still wants to use the site then the additional premium is added to the Contract Sum, and the premium receipt passed to the Employer if requested.

(See English Industrial Estates Corporation v George Wimpey & Co. Ltd, 1972)

Work Not Forming Part of the Contract

2.7   There may be times where the Employer wishes to carry out work on site that is not part of this contract or to employ other people (i.e. Employer's Persons) to carry out such work. For example, an Employer may wish to have his security systems installed by his own security department, which would require his employees to work alongside the Contractor's operatives. As long as one of the correct procedures has been used, the Contractor must allow this work to be executed. The procedures are as follows:

2.7.1   within the Contract Bills there should be information that details the work the Employer wishes to execute. The Contractor must allow the Employer to carry out this work as long as the Contract Bills provide the Contractor with sufficient information to enable him to accommodate the Employer's work and allow him to proceed with his own work in accordance with the contract conditions, or

2.7.2   where the Contract Bills do not provide the necessary information as required above, the Employer may still have such work carried out but he must first obtain the Contractor's consent. The Contractor cannot unreasonably withhold or delay giving his consent.

Note: Any person directly engaged by the Employer would be classified as an Employer's Person; for administrative and insurance purposes, they must not be treated as sub-contractors.

Supply of Documents, Setting Out, etc.
Contract Documents

2.8.1   This clause identifies that it is the Employer who is responsible for holding the Contract Documents (see Definitions clause 1.1 for a list of the items that comprise the Contract Documents). The documents should be available at all reasonable times should the Contractor wish to look at them.

2.8.2   This clause lists the documents the Contractor is entitled to receive from the A/CA, free of charge. The documents listed below are to be supplied to the Contractor immediately after the execution of the contract, unless copies have been previously provided:

2.8.2.1   one copy of the Contract Documents. The documents are to be certified on behalf of the Employer, i.e. initialled or signed to confirm their authenticity;

2.8.2.2   two further copies of the Contract Drawings (see clause 1.1 for definition);

2.8.2.3   two copies of the unpriced bills of quantities.

2.8.3   The Contractor is to keep on site a copy of the Contract Drawings, an unpriced copy of the bill of quantities, CDP Documents (where relevant), descriptive schedules, master programme, setting out drawings and any further drawings and details that may be subsequently issued under clause 2.12, etc. These are to be made available to the A/CA or his representative at all reasonable times.

2.8.4   This clause is written to confirm the confidentiality and copyright in the Contract Documents. The Contractor is not to use the Contract Documents (or schedules and further drawings issued by the A/CA) for any purpose apart from the Works. Similarly the Employer, A/CA or QS are not to make use of the Contractor's rates and prices for any other purpose than this contract.

Construction Information and Contractor's Master Programme

2.9.1   This clause details the information and documentation that must be provided as soon as possible after the execution of the contract, unless copies have been previously provided:

2.9.1.1   Without charge to the Contractor the A/CA is to provide 2 copies of any descriptive schedules, etc. that are necessary to enable the Contractor to carry out the Works, excluding any CDP work where relevant;

2.9.1.2   at no cost to the Employer, the Contractor is to provide the A/CA with 2 copies of a master programme. Whenever the contract period is adjusted, through clause 2.28.1 or by agreement (e.g. a Schedule 2 Quotation) the Contractor is to provide a revised programme within 14 days of notification of the change in the contract period.

(See Moody v Ellis, 1983)

It is important to note that the descriptive schedules and programme cannot create any obligation on the parties greater than the obligations already contained within the Contract Documents. For example, a Contractor is not in breach of contract if he fails to follow his master programme and executes work out of sequence, but he would be in breach of contract if he did not comply with clause 2.4 and failed to complete the Works on or before the Completion Date.

2.9.2   Where the Contractor carries out CDP work he must comply with CDM Regulations 1, 12 and 18,

(*Note*: regulation 1 identifies the relevant CDM legislation that is applicable to this contract, i.e. the Construction (Design and Management) Regulations 2007, which came into force on 6th April 2007. For an explanation of the regulations 12 and 18 refer back to clause 2.2.3.)

and must also provide the A/CA with the following documents, free of charge:

2.9.2.1   two copies of the Contractor's Design Documents (CDD) that are reasonably necessary to explain or expand on the Contractor's Proposals and, if requested by the A/CA, he is also to provide relevant calculations and information that the A/CA considers to be reasonably necessary to explain the Contractor's Proposals;

2.9.2.2   two copies of the levels and setting out dimensions for use on the CDP.

2.9.3   The CDD and other information referred to above must be provided to the A/CA in a timely manner and in accordance with the Contractor's Design Submission Procedure that is set out in Schedule 1 of the contract or in accordance with requirements that may be set out elsewhere in the Contract Documents. It is important that the Contractor is familiar with all the Contract Documents and knows whether there are references to the Contractor's design submission elsewhere. The Contractor must not start any part of the Contractor designed work until the relevant documents have satisfactorily passed through the design submission procedure (see Schedule 1).

Levels and Setting Out of the Works

2.10   This identifies who is responsible for various aspects of the setting out process. The A/CA must provide the Contractor with the appropriate levels and accurately dimensioned drawings to enable the Contractor to set out the Works at ground level. If the Contractor makes a mistake in setting out the Works, he will have to correct it at his own cost, unless the A/CA (with the Employer's agreement) instructs that the errors are not to be corrected, in which case an appropriate deduction may be made from the Contract Sum. The A/CA is not responsible for providing setting out details for CDP work; this is the Contractor's responsibility under clause 2.9.2.2.

Information Release Schedule

2.11   In certain circumstances a Contractor may be provided with an Information Release Schedule which identifies the information that will be provided by the A/CA and the time of its release. From the initial wording of this clause it would appear that the provision of an information release schedule is standard practice. It is in fact an optional clause. The Fifth Recital is used to confirm that the Employer has provided the Contractor with an Information Release Schedule, but footnote [5] then explains that the Employer may delete all reference to the Fifth Recital if he decides that a schedule is not to be provided.

Where an Information Release Schedule is provided, the A/CA is to supply the Contractor with 2 copies of the information at the times stated in the schedule. Obviously, if the Contractor by certain acts or defaults has prevented the A/CA from preparing the information this may remove the A/CA's obligation to comply with the scheduled dates. For example, the Contractor may have failed to provide some design documentation which makes it impossible for the A/CA to prepare a number of associated working drawings.

The Employer and Contractor may agree to amend the times, stated in the schedule, for the supply of information. For example, for any number of reasons the programming of the Works may be altered from what was originally intended with the result that the dates contained in the original Information Release Schedule are no longer appropriate. Either party may suggest an alteration to the timing of the schedule, in which case the other party may not unreasonably withhold their agreement.

Failure by the A/CA to comply with the times given in the schedule may lead to a request, from the Contractor, for an extension of time (clause 2.29.6) and the submission of a claim for loss and/or expense (clause 4.24.5).

Further Drawings, Details and Instructions

2.12.1   This sub-clause sets out the A/CA's obligation to provide the Contractor with further information, such as drawings and details which expand upon the Contract Drawings, and to give instructions which will allow the Contractor to complete the Works in accordance with the contract. Where the drawings and details have not been included in an Information Release Schedule the A/CA is still obliged to provide the Contractor with the necessary information to allow him to complete the Works.

2.12.2   The drawings, details and instructions must be given to the Contractor at a reasonable time in accordance with the progress of the Works and the contract Completion Date. If it is apparent that the Works are ahead of schedule and may reach practical completion before the contract Completion Date, the A/CA is not obliged to keep pace with the accelerated programme, his obligation is to provide the information in a timely manner that allows the Contractor to complete the Works by the contract Completion Date.

(For further information concerning the A/CA's obligation to provide information on an accelerated project see Glenlion Construction Ltd v Guinness Trust, 1987)

2.12.3   If the Contractor considers the A/CA may not be aware of the time by which he needs further information or instructions then, if possible, he should give the A/CA sufficient advance notice of his requirements to enable the A/CA to fulfil his obligations under this clause.

It is important for the Employer to ensure that clauses 2.11 and 2.12 are complied with because if there is a failure to provide drawings and details in a timely manner the Contractor may be entitled to an extension of time (clause 2.29.6) and may be able to submit a claim for loss and/or expense (clause 4.24.5).

Errors, Discrepancies and Divergences
Preparation of Contract Bills and Employer's Requirements

2.13.1   The Contract Bills are to be prepared in accordance with the Standard Method of Measurement (SMM). The Quantity Surveyor may use alternative measurement rules, but every item that is not measured in accordance with the SMM must be clearly identified in the Contract Bills. The same rules apply where an addendum bill is prepared to enable the Contractor to provide a Schedule 2 Quote.

The Standard Method of Measurement is described in clause 1.1 as being the 7th Edition produced by the Royal Institution of Chartered Surveyors and the Construction Confederation, and to be the version current at the contract Base Date. If a different version has been used, this must be clearly stated in the Contract Bills.

2.13.2 Where a Contractor is required to carry out CDP work this sub-clause confirms that the Contractor has no responsibility for the contents of the Employer's Requirements or for checking the design that may be included in the Employer's Requirements. However, there is a reminder that the Contractor must still comply with clause 2.17, i.e. it is the Contractor's responsibility to ensure that the CDP Documents comply with the relevant Statutory Requirements.

Contract Bills and CDP Documents – Errors and Inadequacy

2.14.1 If there is an error in the Contract Bills or an addendum bill for a Schedule 2 Quote (e.g. not measured in accordance with the stated rules, an incorrect description, the wrong quantity or an item has been omitted), then it is to be corrected. Any error in the Contract Bills will not vitiate the contract, i.e. the contract will not be terminated because of this error. An error is to be corrected.

Specific advice is provided where an error occurs in the Contract Bills when describing a 'Provisional Sum for defined work' (look at SMM7 General Rules 10.3 to see what criteria are used to determine whether work is classified as defined or undefined, more importantly see what the consequences are regarding programming and the cost of preliminaries). Where a description for defined work is inadequate, the description is to be amended so that it complies with the SMM7 rules.

2.14.2 If there is an error or inadequacy in any design incorporated into the Employer's Requirements, then the Employer's Requirements are to be amended so as to remove the problem. However, if the Contractor's Proposals have already resolved the problem, then no action is required.

2.14.3 Any corrections or alterations carried out under clause 2.14 will be treated *as though they were a variation*, i.e. the Contractor is entitled to recover associated costs in accordance with the valuation of variation, etc. but there is no need for the A/CA to actually issue a variation instruction. Again there is a reminder that alterations required because of a divergence with the Employer's Requirements and Statutory Requirements are normally the Contractor's responsibility (see clause 2.17.2.1)

2.14.4 If there is an error in the Contractor's Proposals or the CDP Analysis (e.g. an error in description, a wrong quantity or items omitted) it is to be corrected at no cost to the Employer. If the correction of the error results in the A/CA having to issue a variation to the non-CDP work, this again is at no cost to the Employer.

Notification of Discrepancies etc.

2.15 If the Contractor discovers an error or inadequacy, as detailed in clause 2.14, he is to immediately write to the A/CA giving details of the problem. Once the A/CA has received the notification it is his duty to resolve the error, etc. through the issue of an appropriate instruction.

Occasionally discrepancies or divergences may occur between some of the Contract Documents or subsequent drawings and schedules issued by the A/CA or the CDP Documents, for instance there could be conflicting information provided by two Contract Drawings or between a drawing and the Contract Bills. The documents and drawings that may be considered under this clause are listed below:

2.15.1 the Contract Drawings, these are the drawings referred to and identified in the Third Recital;

2.15.2 the Contract Bills, these are the bills referred to in the Second Recital;

2.15.3 any instruction issued by the A/CA; any instructions that are variations are excluded from this list – the reason being that a variation by its very nature will be a change from the Contract Documents;

2.15.4 any drawing or document issued by the A/CA under clauses 2.9 to 2.12 – this refers to further details that the A/CA may provide during the progress of the project, e.g. descriptive schedules, setting out drawings, further drawings and details;

2.15.5 CDP Documents – these will only exist where the Contractor has been required to carry out CDP work.

If the Contractor becomes aware of a discrepancy or divergence between any of the above documents then again he is required to immediately write to the A/CA giving details of the problem, once the A/CA has received the notification it is his duty to resolve the discrepancy etc. through the issue of an appropriate instruction.

With reference to clause 2.15 it is a generally accepted principle that a Contractor is not required to comb through the documents produced by the Employer or his advisers looking for any errors or discrepancies, but if and when he does notice a discrepancy he must immediately inform the A/CA through a written notice.

(See London Borough of Merton v Stanley Hugh Leach Ltd, 1985;

Equitable Debenture Assets Corporation Ltd v William Moss Group Ltd and Others, 1984;

Oxford University Press v John Stedman Design Group and Others, 1990;

Plant Construction plc v Clive Adams Associates and JMH Construction Services, 1999)

Discrepancies in CDP Documents

2.16.1 Where a discrepancy or divergence, as identified in clause 2.15, occurs within or between the CDP Documents (see Definition in clause 1.1) other than the Employer's Requirements, the Contractor, as well as sending a notice of divergence to the A/CA, must also provide a statement of how he intends to resolve the issue. Where possible the statement is to be sent with the notice, or as soon as practicable after the issue of the notice. The A/CA's obligation to issue an instruction under 2.15 is suspended until the Contractor's statement is received. When the instruction is issued it will be at no cost to the Employer. It is important to note that discrepancies, etc. within the Employer's Requirements are not to be dealt with in this clause and should be dealt with in accordance with clause 2.16.2.

2.16.2   Where a discrepancy notified in clause 2.15 occurs within the Employer's Requirements (this also includes variations that may affect the Employer's Requirements) the work will be executed in accordance with the Contractor's Proposal, as long as there is no conflict with Statutory Requirements. Obviously in this instance there will be no adjustment to the Contract Sum. Where the Contractor's Proposals do not provide a solution to the discrepancy, the Contractor must provide the A/CA with a written statement detailing his proposed solution. The A/CA may accept the Contractor's solution or produce his own solution. His decision must be given to the Contractor in writing and is to be treated as a variation, i.e. a potential cost to the Employer.

Divergences from Statutory Requirements

2.17.1   If either the Contractor or A/CA finds there is a divergence between the Statutory Requirements and any of the following:

2.17.1.1   any of the documents listed in clause 2.15.1 to 2.15.5; or

2.17.1.2   any variation instruction that has been issued under clause 3.14,

then whoever finds the divergence must immediately give a written notice to the other identifying the divergence. Where there is a divergence between the Statutory Requirements and any of the CDP Documents, the Contractor must write to the A/CA with a proposal as to how the divergence may be remedied.

2.17.2   Upon receiving the Contractor's notification of a divergence, or from the date the A/CA discovers a divergence, he must issue an instruction within 7 days to resolve the problem. However, a different time period is to apply where the divergence relates to the CDP Documents. In this instance the A/CA must issue an instruction within 14 days from when the Contractor submitted his proposed amendment to the CDP Documents. The effect of the instruction is as follows:

2.17.2.1   where the divergence was between the Statutory Requirements and any of the CDP Documents, then the Contractor must comply with the instruction at no cost to the Employer. The exception to this is where the divergence came into existence because the Statutory Requirements were changed after the Base Date and as a result the Contractor is now required to make changes to the CDP work. In this case any alteration to the CDP work will be treated as a variation, i.e. at the Employer's expense;

2.17.2.2 in all other cases (i.e. where CDP Documents are not involved) where the instruction requires the Work to be varied, it will be treated as a variation, again this will be at the Employer's expense.

2.17.3 As long as the Contractor has notified the A/CA of any divergence he has discovered between the Statutory Requirements and the Contract Documents, variations, drawings etc., then the Contractor will not be liable to the Employer if the Works, as specified, do not comply with the Statutory Requirements. The exception to this is where there is a divergence involving CDP Works, in which case the Contractor is liable to the Employer for the non-compliance.

A Contractor must accept responsibility for the information contained within his contractor' proposals, but the general principle is that a Contractor is not obliged to comb through all the other contract documentation to identify any divergences with Statutory Requirements. However, if and when a Contractor does become aware of a divergence he is under an obligation to notify the A/CA.

(See London Borough of Merton v Stanley Hugh Leach Ltd, 1985)

Emergency Compliance with Statutory Requirements

2.18.1 If an emergency situation arises in relation to Statutory Requirements (i.e. imminent danger to persons or property) which requires the Contractor to act before he receives an instruction under clause 2.17.2, the Contractor may immediately carry out additional work to avert the emergency. It is important for the Contractor to note that he should carry out only the minimum amount of work reasonably necessary to ensure he has complied with the Statutory Requirements in relation to the emergency.

2.18.2 The Contractor must promptly inform the A/CA of the emergency, and what steps have been taken to resolve the problems.

2.18.3 As long as the emergency arose because of a conflict between the Statutory Requirements and the Contract Documents, etc., and as long as the Contractor notified the A/CA as required by clause 2.18.2 then the cost of any work will be borne by the Employer. The work by the Contractor will be deemed to be covered by a variation instruction, i.e. there is no need for the A/CA to actually issue an instruction. The work will be valued in accordance with clause 5.2. If the emergency arose because of a conflict with CDP Documents then the costs will remain with the Contractor.

CDP Design Work
Design Liabilities and Limitation

This section explains the Contractor's design liability where he has been made responsible for an element of the design through the Contractor's Designed Portion.

2.19.1 Where a Contractor has provided design details through the CDP procedure, his liability to the Employer for any failure in the design is the same as if the design work had been carried out by an architect. This is advantageous to the Contractor as it reduces his potential liability to the Employer. Where a Contractor takes on the responsibility for both the design and construction of the Works it is normally implied that the design shall be 'fit for its purpose'. This is a very onerous liability as it makes the Contractor potentially responsible for any flaw in the design that has an adverse effect on the usability of the building. By contrast, where an architect or other professional designer is engaged by an Employer, it is normally implied that his design liability is limited to exercising 'reasonable skill and care'. If an architect's design proves to be faulty he would only be liable to the Employer if it could be shown that the standard of design fell below that of a normal competent architect.

If the Contractor's design relates to work that would not normally be carried out by an architect, e.g. mechanical and electrical services, then the Contractor's liability for the design is the same as if the design had been carried out by a suitable competent professional designer (e.g. a consulting engineer) engaged by the Employer. Again the implied liability for such a professional designer would be one of reasonable skill and care.

2.19.2 This is a reminder that, where a designer carries out work in relation to a dwelling (i.e. a house, apartment, etc.) the work will be covered by the Defective Premises Act, i.e. there is a responsibility *'to see that the work which he takes on is done in a workmanlike or, as the case may be, professional manner, with proper materials and so that as regards that work the dwelling will be fit for habitation when completed'.*[1]

2.19.3 Where the Contractor is engaged in non-domestic work, his design liability under clause 2.19.1 includes loss of use, loss of profit or other consequential loss. The extent of the Contractor's liability may be capped by inserting a financial limit in the Contract Particulars. If this section in the Contract Particulars is left blank, it must be assumed that the Contractor's liability is open-ended. It is pointed out that, where a limit is provided it has no relevance to the operation of the liquidated damages provisions in clause 2.32. For example, where a Contractor is late in completing the Works and is incurring liquidated damages, he cannot claim that his liquidated damages are capped at a limit stated under clause 2.19.3. Similarly, if an Employer is claiming loss of profit because of a design fault, the Contractor cannot claim that previously deducted liquidated damages should be included in determining when the limit has been reached.

Errors and Failures – Other Consequences

This section deals with the issue of delays and suspension of the Works caused by the Contractor's design.

2.20 This clause identifies that the Contractor:

> is not entitled to claim an extension of time;
>
> is not entitled to claim for loss and/or expense under the provisions of clause 4.23;
>
> is not entitled to exercise his right of termination under clause 8.9.2 (i.e. following the suspension of Works);

where the Works have been delayed, disrupted or suspended by any of the following:

2.20.1 an error, divergence, omission or discrepancy in the Contractor's Proposals, or in the information provided by the Contractor under clause 2.9.2 (i.e. this may include information such as the Contractor's Design Documents along with any associated calculations and information, as well as dimensions and levels provided by the Contractor for setting out the CDP work);

Any failure of the Contractor to comply with CDM Regulations 11, 12 and 18 (see clause 2.2.3 for an explanation) when completing the Contractor's Design Documents.

2.20.2 failure by the Contractor to provide Contractor's Design Documents, calculations, information, etc. on time. That is where the Contractor has failed to comply with one of the following requirements:

2.20.2.1 clause 2.9.3 which sets out the timing for the delivery of these documents, or

2.20.1.2 a written request from the A/CA specifying the documents or information he requires, and the date for receipt. Any such request from the A/CA must be reasonable in relation to the progress of the Works and the Completion Date. This means that where the Contractor is working ahead of schedule the A/CA is not obliged to keep pace with the Contractor's accelerated progress and may request design documents at a time that is reasonable with reference to the Completion Date.

(See Glenlion Construction Ltd v Guinness Trust, 1987 for a more detailed explanation)

Fees, Royalties and Patent Rights
Fees or Charges Legally Demandable

2.21 It is the Contractor's duty to pay for any fees that are legally demanded by any Statutory Requirement in connection with the Works. Where the fee is stated as a Provisional Sum (for example, a Provisional Sum for connection to a water main) the payment is dealt with under clause 4.3, i.e. the Provisional Sum is deducted from the Contract Sum and the actual fee is added back. In other instances the fee or charge will be added to the Contract Sum, unless it has been described in the Contract Bills (so the Contractor was able to include the price in his tender) or where the fee arises in connection with CDP work, when it is deemed to be included in the Contractor's tender.

Royalties and Patent Rights – Contractor Indemnity

2.22   If the Contract Bills or Employer's Requirements describe works that require the use of patent materials, processes, etc. the Contractor is to allow for the cost of any ensuing royalties, etc. If the Contractor fails to comply with any requirements relating to a patent process, and as a result of such infringement proceedings are brought against the Employer, then the Contractor is liable to the Employer for any resultant costs and damages. A Contractor would be advised to check any limitations or obligations regarding the use of a patent process at the tender stage.

Patent Rights – Instructions

2.23   Where the Contractor is required to use a patent process, etc. as a result of an A/CA's instruction then any royalties or costs associated with an infringement will be added to the Contract Sum.

Unfixed Materials and Goods – Property, Risk etc.
Materials and Goods – On Site

2.24   Materials that are on site are not to be removed from the site without the A/CA's written consent. The A/CA may not withhold his consent unreasonably. Once the Employer has paid for these materials through an Interim Certificate they are to become his property, although the Contractor is still responsible for any loss or damage to the materials. The Contractor's liability for loss and damage is reduced where the Employer has taken out the All Risks Insurance under Insurance Option B or C, i.e. materials damaged by fire, etc. would be covered by the Employer's insurance. The Contractor would remain liable for any damage or loss not covered by the All Risks Insurance.

(See Aluminium Industrie Vaassen BV and Romalpa Aluminium Ltd, 1976; Dawber Williamson v Humberside County Council, 1979)

Materials and Goods – Off Site

2.25   Where the Employer has paid the Contractor for 'listed-items' (see Definitions clause 1.1) held off-site, these materials become the property of the Employer. The Contractor is not to allow the materials to be removed from their storage unless it is to transport them to site. The Contractor remains liable to pay any storage costs and is responsible for any loss or damage to the materials whilst they are off-site. He is to insure the materials against loss and damage through a 'Specified Perils' policy' (see clause 4.17.2.2). Once the materials are delivered to site the Contractor's obligations are covered by clause 2.24.

Adjustment of Completion Date
Related Definitions and Interpretations

2.26   This section provides definitions and explanations of some of the terms used in clauses 2.27 to 2.29:

2.26.1 Where a reference is made to 'delay' or 'extension of time' it is deemed to also include, where relevant, further delay or further extension of time.

2.26.2 A Pre-agreed Adjustment comes into existence through a Schedule 2 Quotation. Where a Contractor provides a Schedule 2 Quotation he is to give notice of any 'adjustment of time' that may be required to carry out the Schedule 2 Quotation work (see Schedule 2 paragraph 2.2); he may require extra time, or in some circumstances, the contract period may be shortened. The A/CA is to confirm any adjustment of time when he issues his acceptance of the quotation (see Schedule 2 para.3.2.3) and this is referred to as a Pre-agreed Adjustment.

2.26.3 The term 'Relevant Omission' refers to work or obligations that have been omitted through either a variation instruction under clause 3.14 or an instruction relating to a Defined Provisional sum issued under clause 3.16.

Notice by Contractor of Delay to Progress

This section sets out the Contractor's obligation to keep the A/CA informed about delays or potential delays in the progress of the work.

2.27.1 The Contractor must promptly notify the A/CA whenever it becomes reasonably apparent that the Works is being delayed, or is likely to be delayed. The written notice must provide details of the delay, the cause(s) of the delay and any Relevant Event that may have caused or contributed to the delay. It is important to note that the Contractor is required to give notice of potential future delays of which he may be aware, and not to wait until the delay has actually occurred before issuing his notice.

(See London Borough of Merton v Stanley Hugh Leach Ltd, 1985)

2.27.2 Where the Contractor has identified a Relevant Event(s) in the above notice he should, where possible, include particulars (i.e. information, details) about the expected effect of each event and an estimate of the overall delay. If it is not practicable to provide these particulars in the notice it must be provided to the A/CA, in writing, as soon as possible after the submission of the notice. In providing this information the Contractor should be aware of the need to provide an estimate of how each Relevant Event may delay the Works beyond the Completion Date, ignoring the fact that one delay may run concurrently with a delay caused by another Relevant Event. For example, three Relevant Events may have been notified, each causing a delay of 3 weeks, i.e. 9 weeks in total; however, because two of the events are concurrent the overall delay to the Works is only 6 weeks.

2.27.3 The Contractor is to promptly notify the A/CA of any changes in the particulars or the estimate of the delay previously provided. This must be given in writing. The Contractor is also obliged to provide further information that the A/CA may reasonably request.

Fixing Completion Date

This section details the procedures to be followed by the A/CA in reviewing whether or not the completion date is to be adjusted.

2.28.1 Where an A/CA has received a notice and particulars of a delay under clause 2.27 he must determine whether any of the events which the Contractor has identified as having caused a delay is a Relevant Event (see list in clause 2.29). The A/CA must also satisfy himself that the Works, or Section, is likely to be delayed by the identified Relevant Event(s). Where the A/CA is satisfied that these issues have been properly dealt with he is to allow an extension of time and set a new Completion Date for the Works or a Section. The extension awarded by the A/CA can only be an estimate based on the information available at that time, but it must be seen to be 'fair and reasonable'. Obviously, if the A/CA is of the opinion that the Works will not be delayed beyond the Completion Date, then no extension of time will be awarded. The A/CA is advised that there may be clauses elsewhere in the contract that may modify his right to award an extension of time under clause 2.28.1 (e.g. see clause 2.20).

(See London Borough of Merton v Hugh Stanley Leach Ltd, 1985)

2.28.2 The A/CA must notify the Contractor (in writing) whether or not an extension of time is to be granted. Where an extension is granted the A/CA should, if reasonably possible, provide the new Completion Date within 12 weeks from receipt of the Contractor's notice or receipt of the required particulars (i.e. as required under clause 2.27.2), whichever is the later. If there is less than 12 weeks between receipt of the Contractor's notice, particulars or estimate and the Completion Date then the A/CA should try and give notice of the new Completion Date, before the current Completion Date is reached. An important proviso is that the above time periods imposed on the A/CA are conditional upon the Contractor having provided the required particulars, etc. to allow the A/CA to reach a decision.

Where the A/CA considers that it is not fair and reasonable to set a new Completion Date, he should notify the Contractor in accordance with the above timescales.

Note: the extension of time awarded at this stage is only an estimate and the A/CA may review and increase the time allowed after practical completion (see clause 2.28.5.1). However, the A/CA may not subsequently reduce an extension of time (see clause 2.28.5.2) unless there is a corresponding reduction in work executed.

2.28.3 In the above notice the A/CA must inform the Contractor of the extension of time he has allocated to each Relevant Event. Similarly, where the A/CA has reduced the contract period under clause 2.28.4 or 2.28.5.2 he must identify the time reduction allocated to each Relevant Omission (see clause 2.26.3).

2.28.4   This clause allows the A/CA to reduce a previously issued extension of time. After the A/CA has issued an extension of time (either under clause 2.28 or through the acceptance of a Schedule 2 Quotation) the A/CA may subsequently give the Contractor a written notice to the effect that he is setting an earlier Completion Date. The A/CA's action has to be fair and reasonable and must be a direct result of work having been omitted since the last extension of time was awarded, i.e. a Relevant Omission. In the notice the A/CA must give the information required in clause 2.28.3.

The A/CA must be aware that his decision to reduce the contract period must not conflict with clauses 2.28.6.3 and 2.28.6.4 (see below for further details).

2.28.5   This clause allows the A/CA to review previous alterations to the contract period and give the Contractor a written notice fixing the Completion Date for the project. The timing of this notice is dependent upon the progress of the Works (Section). Where the Works (or Section) are in delay, i.e. they have not yet reached practical completion but the current Completion Date has been passed, then the A/CA *may* issue the notice. However, where practical completion of the Works (or Section) has been achieved, the A/CA must issue the notice within 12 weeks from that date. It is likely that most A/CAs would wait until practical completion before commencing this exercise. In this notice the A/CA must provide the detail required by clause 2.28.3 and notify the Contractor of his decision, i.e.:

2.28.5.1   that a later Completion Date is to now apply. This Completion Date should be fair and reasonable in the light of any Relevant Events that have occurred. The A/CA is expected to review all Relevant Events associated with the project. This would mean considering previously reviewed Relevant Events and newly identified Relevant Events. The A/CA should not limit himself to those Relevant Events notified by the Contractor but should completely review the project to satisfy himself that all Relevant Events have been identified; or

2.28.5.2   that an earlier Completion Date is to apply where this would be reasonable (and in accordance with clauses 2.28.6.3 and 2.28.6.4) in the light of Relevant Omissions, for which instructions were issued after the last occasion when a new Completion Date was fixed; or

2.28.5.3   that the previously fixed Completion Date is to remain.

2.28.6   This section imposes certain obligations on the Contractor with regard to delays and potential delays and advises the A/CA about the limits regarding his powers to reduce the contract period.

2.28.6.1   The Contractor should always use his 'best endeavours' to prevent the work being delayed beyond the completion or revised Completion Date, regardless of who is responsible for the delay. The meaning of 'best endeavours' is subject to debate; a popular opinion is that a Contractor should consider all reasonable means of increasing his efficiency by reviewing his resources to overcome or moderate the effect of the delay, but this should not require any additional expenditure, e.g. night time working, hiring of specialist plant, etc.;

2.28.6.2   Steps taken by the Contractor should be to the reasonable satisfaction of the A/CA, therefore the A/CA may be able to influence the actions of the Contractor;

2.28.6.3   The A/CA when operating clause 2.28.4 or 2.28.5.2 cannot fix a Completion Date earlier than the Date for Completion originally stated in the Contract Particulars;

2.28.6.4   The A/CA cannot reduce an extension of time awarded under a Pre-agreed Adjustment, unless the Schedule 2 Quotation work (or 'other work' referred to in the Quotation) is itself the subject of a Relevant Omission. As an example, the original contract completion for a certain project was 52 weeks which was extended by 10 weeks following the acceptance of a Schedule 2 Quotation (8 weeks extension) and a claim for exceptionally adverse weather (2 weeks extension). At a later date an instruction is issued to omit a large amount of external works from the contract (i.e. a Relevant Omission) which reduces the Contractor's programme by 3 weeks. However, the A/CA may only subsequently reduce the contract period by 2 weeks, i.e. the extension of time granted for exceptionally adverse weather conditions. It is not possible to reduce any of the 8 weeks extension granted for the Schedule 2 Quotation as the Relevant Omission did not omit any of the work being carried out under the Schedule 2 Quotation.

Relevant Events

2.29   This is an important clause that lists the various events that may entitle the Contractor to request an adjustment to the Completion Date. The Relevant Events are as follows:

2.29.1   variations, which will also include events or instructions that are '*to be treated as a variation*' (e.g. see clause 2.14.3) or '*as requiring a variation*' (e.g. see clause 2.17.2.2);

2.29.2   instructions issued by the A/CA in connection with the following:

2.29.2.1   clause 2.15, instructions issued in response to discrepancies discovered within the Contract Documents;

clause 3.15, an instruction to postpone the Works;

clause 3.16, instructions as to the expenditure of Provisional Sums but excluding instructions for the expenditure of defined Provisional Sums. Defined Provisional Sums are excluded because the Contractor is deemed to have made an allowance for this work at tender stage and cannot normally request an extension of time for the work that is carried out;

2.29.2.2   clause 3.17 and 3.18.4, instructions for the opening up and testing of the Works and making good. A Contractor is not able to claim under this heading if the tests show that the work is defective;

2.29.3   where the Employer delays giving the Contractor possession of the site. This only applies where clause 2.5 is operative and the deferment is in accordance with this clause, e.g. the Employer has not deferred possession of the site beyond the maximum period allowed in the Contract Particulars. If this were to happen the Employer would be in breach of contract and the A/CA, technically, could not issue an extension of time to compensate for the Employer's breach;

(See Rapid Building Group Ltd v Ealing Family Housing Association Ltd, 1984)

2.29.4   where the Contractor has carried out work covered by an Approximate Quantity in the bill of quantities and the Approximate Quantity proves not to be a reasonably accurate forecast of the work to be executed. For instance, an Approximate Quantity may have grossly underestimated the amount of work actually required, causing delay to the Contractor, as a result of executing this additional work;

2.29.5   delay as a result of the Contractor exercising his right to suspend carrying out the Works under clause 4.14, i.e. suspension of the Works as a consequence of the Employer's failure to pay the Contractor amounts that were properly due;

2.29.6   delay caused by a Statutory Undertaker when carrying out work in accordance with its statutory duties, or the failure to carry out these works;

2.29.7   exceptionally adverse weather conditions; the weather conditions have to be exceptional for the time of year and have an adverse effect on the work, e.g. a six-week heat wave during the summer could come under the category of 'exceptional' if records from the local meteorology centre confirm that this is a weather pattern that rarely occurs. If the prolonged heat wave has an adverse effect on any of the Contractor's operations (e.g. asphalt tanking work may have to be delayed) a claim for an extension of time may be allowed;

2.29.8   Works or Site Materials damaged by one or more of the Specified Perils, e.g. fire, flood, etc. (see clause 6.8 for a complete definition of Specified Perils);

2.29.9   civil commotion, or the use or threat of terrorism, or the action(s) taken by the appropriate authorities in dealing with such terrorism. For example, the Contractor may be refused access to the site by the authorities because of a terrorist threat or incident;

2.29.10   strikes, etc. affecting trades employed on site, or engaged in work off-site, or persons engaged in design work for CDP;

2.29.11   where, after the Base Date, the UK government uses its statutory powers, which action directly affects the execution of the Works, e.g. restricting the Contractor's ability to obtain labour, materials, goods or fuel which are essential for carrying out the Works. For example, on 13 December 1973 the British Government announced stringent measures to conserve electricity; commercial organisations were only able to use electricity on three consecutive days in each working week;

2.29.12   force majeure; there is no accepted legal definition of this phrase but it is generally accepted as referring to exceptional matters beyond the control of the parties.

Practical Completion, Lateness and Liquidated Damages

This section relates to some of the procedures relating to the completion of the project and the consequences of late completion.

Practical Completion and Certificates

2.30   Once the Works has reached practical completion the A/CA is to immediately issue a Practical Completion Certificate. Whether or not the work has reached practical completion is down to the opinion of the A/CA; unfortunately there is no definitive legal definition of the term 'practical completion', therefore it is open to a wide degree of interpretation. Prior to issuing this certificate the A/CA must be satisfied that the Contractor has provided the Employer with the required as-built drawings for any relevant CDP work and complied with any requests for information from the CDM Co-ordinator which he reasonably requires so that the health and safety file may be prepared. Practical completion will take place on the day stated by the A/CA in the certificate, which will not necessarily be the same as the date of issue of the certificate.

Where a project has been broken up into Sections, then the A/CA will issue Section Completion Certificates, and the same procedures as above will apply.

(See City of Westminster v J Jarvis,1970;

H W Nevill (Sunblest) Ltd v Wm Press & Sons Ltd, 1981;

Emson Eastern Limited (In Receivership) v EME Developments Ltd, 1991)

Non-Completion Certificate

This clause identifies the action to be taken by the A/CA where the Contractor fails to complete the Works, or a Section, by the relevant completion date.

2.31   If the Contractor has not completed the Works (or a Section) by the Completion Date, the A/CA is required to issue a Non-Completion Certificate. If the A/CA grants the Contractor an extension of time after this event, the Non-Completion Certificate becomes invalid because a new Completion Date has been created.

Therefore, if the Contractor subsequently fails to complete the Works by the revised Completion Date, the A/CA must issue another Non-Completion Certificate.

(See A Bell & Son Ltd v CBF Residential Care and Housing Association, 1989)

Payment or Allowance of Liquidated Damages

This clause informs the Contractor of the financial implications of failing to complete the works (or sections) within the time allowed. The clause also sets out the procedures to be followed by the Employer if he is to successfully recover liquidated damages from the Contractor.

(See Temloc Ltd v Erril Properties Ltd, 1987; Bramall & Ogden Ltd v Sheffield City Council, 1983)

2.32.1   Before the Employer is entitled to recover liquidated damages, it is a strict requirement that the A/CA must have issued a valid Non-Completion Certificate. Also the Employer must have written to the Contractor informing him that liquidated damages may be required or may be withheld or deducted from monies due to the Contractor. The Employer's notification must be issued before the date of the Final Certificate.

As long as the above procedures have been complied with, the Employer may subsequently give the Contractor a written notice specifying the amount of liquidated damages being claimed. The notice must be issued no later than five days before the final payment is due under the Final Certificate (see clause 4.15.4). The terms of the notice are set out in clause 2.32.2, as detailed below.

2.32.2   The notice from the Employer to the Contractor should state the Employer's intention to recover liquidated damages for the period between the Completion Date and the date of practical completion for the Works or relevant Section. The notice should identify whether the Employer is seeking to recover liquidated damages at the rate stated in the Contract Particulars or at a lesser rate – in which case the lesser rate should be clearly stated. The Employer has two ways of recovering the money: he may require the Contractor to pay the amount to him as a debt, or the Employer may withhold or deduct the liquidated damages from monies due to the Contractor. Where the Employer follows the latter approach it is essential that the Contractor is issued with a proper withholding notice. If all the necessary information is included in the clause 2.32.2 notice this would satisfy the requirements of a withholding notice, otherwise a separate withholding notice will have to be issued.

It is important to realise that in the instance where the liquidated damages are calculated at the rate stated in the Contract Particulars this rate may need to be adjusted in the light of any partial possession.

As long as the necessary documentation is in place, the Employer may start recovering liquidated damages once the Contractor has failed to meet the Completion Date; the Employer does not have to wait until practical completion or the Final Certificate. The Employer could, for example, withhold liquidated damages from future Interim Certificates.

2.32.3 Where the Employer has recovered liquidated damages but the A/CA subsequently fixes a later Completion Date, this means that the liquidated damages sum will have to be recalculated and the difference returned to the Contractor. If, for example, the Employer has withheld 60 days of liquidated damages from the Contractor when the A/CA extends the Completion Date by 10 days, the Employer will have to allow or repay to the Contractor 10 days of liquidated damages.

2.32.4 Once the Employer has given a written notice of his intention to deduct liquidated damages, that notice remains valid even though the relevant Non-Completion Certificate may have been cancelled through the setting of a later Completion Date (clause 2.31), unless the Employer writes to inform the Contractor that he has withdrawn the notice.

(See A Bell & Son Ltd v CBF Residential Care and Housing Association, 1989)

Partial Possession by Employer

This is a procedure by which the Employer may take possession of part of the Works that has reached completion. Under clause 2.4 the Contractor is given full possession of the site. During the progress of the project, however, certain discrete sections of the Works may be completed by the Contractor and the Employer may wish to take over those sections, i.e. take possession of them.

Contractor's Consent

2.33 With the consent of the Contractor (this may not be unreasonably delayed or withheld) the Employer may take possession of part of the Works before practical completion or Section completion. The A/CA must subsequently issue a written statement that identifies the part(s) of the work taken over by the Employer and the date when this occurred. For the purposes of this procedure these events are referred to as the Relevant Part and Relevant Date respectively.

Practical Completion Date

2.34 The part(s) taken over by the Employer is deemed to have reached practical completion and therefore the rules in clause 2.38 relating to defects, etc. will apply. Also the rules relating to Retention (clause 4.20.3) will apply, resulting in a release of half the Retention on the value of the work taken over by the Employer.

Defects etc. – Relevant Part

2.35 When the Contractor has remedied all the notified defects for the part taken over by the Employer, the A/CA is to issue a certificate to confirm the fact. This allows the final release of Retention for this part of the work (see clause 4.20.3).

Insurance – Relevant Part

2.36   On new Works, where either the Contractor or Employer is responsible for taking out the All Risks Insurance (under either Option A or Option B), the obligation to maintain this insurance ceases (from the date on the A/CA's statement) for the portion of the Works taken over. Unless the Employer arranges further insurance, the Works taken over will be largely uninsured. If the main Works comprise alterations or extensions to an existing building then the Works have to be insured by the Employer under clause Option C.2. This insurance will terminate (for that part of the Works) on the date the Employer takes over part of the Works, but in this instance the Employer must now add that part of the Works to his Option 22C.1 insurance, which provides cover for the Employer's existing building.

Liquidated Damages – Relevant Part

2.37   As a result of partial possession the amount of liquidated damages the Employer may claim as a result of the Contractor's late completion is reduced. The amount of liquidated damages is stated in the Contract Particulars, and that figure is reduced by the same proportion as the value of the part taken over by the Employer compared with the Contract Sum. For example, an Employer, on a £2m project, takes over part of the Works valued at £200,000. Liquidated damages have been inserted in the Contract Particulars at £1,000 per day, but from the date of partial possession the liquidated damages for the remaining Works will be deemed reduced by 10%, i.e. damages will now be set at £900 per day.

Defects
Schedules and Instructions

This section sets out the procedures for the Rectification Period, a period of time when the Contractor may be required to remedy defective works. The Rectification Period runs from the date of practical completion or Section completion (see Definitions clause 1.1 and contract particulars). The length of time for which the Rectification Period will run is to be stated in the contract particulars – if this section is left blank a default period of 6 months will apply.

2.38.1   During the Rectification Period the A/CA may issue an instruction providing the Contractor with a schedule of defects (i.e. workmanship or materials that do not comply with the specification or fail to comply with obligations relating to CDP). The last day by which the A/CA may issue the schedule is 14 days from the end of the Rectification Period.

2.38.2   In addition to the provisions above, the A/CA is also able to issue instructions for the remedying of defects throughout the Rectification Period, but this right ceases the moment the schedule of defects is issued or when more than 14 days has passed since the end of the Rectification Period.

The Contractor must remedy the defects at his own expense and within a reasonable time from receipt of the notification from the A/CA. Alternatively, with the Employer's consent, the A/CA may instruct that certain defects are not to be remedied, in which case an appropriate deduction is to be made to the contract sum. For example, a minor defect may exist such as the installation of inferior door ironmongery; the Employer may prefer to retain this ironmongery instead of facing considerable disruption if the ironmongery were to be replaced with the type specified in the contract documents.

(See Pearce & High Ltd v Baxter and Anor, 1999)

Certificate of Making Good

2.39   When the A/CA considers that all defects have been made good, where required, he is to issue a Certificate of Making Good. In the certificate the A/CA is to name the day on which he was satisfied that all defects had been made good. A certificate may be issued for a Section(s) of the Works or for the whole Works.

Contractor's Design Documents
As-built Drawings

This details the documentation the Contractor may have to provide to the employer where the contractor has been required to carry out a portion of the design for the project.

2.40   Where the Contractor has undertaken CDP work it may be a requirement of the Contract Documents that the Contractor must provide the Employer with documents showing the Contractor's Designed Portion as it has been built, along with information relating to the maintenance and operation of that portion or any installations contained within the portion (e.g. lift installations, heating systems, air-conditioning, etc.). Alternatively, the Employer may make a reasonable request for the supply of such information. In either instance the Contractor is to provide the documentation before practical completion and at no charge to the Employer. It is important to note that compliance with this clause does not remove or diminish the Contractor's obligations to provide the relevant health and safety information required under clause 3.25.

Copyright and Use

2.41.1   The Contractor retains the copyright in the Contractor's Design Documents. This prevents the Employer or A/CA from using any part of the Contractor's design on other projects, unless they have the Contractor's consent. However, the Contractor's copyright in the CDD does not apply to designs, drawings, etc. provided by the Employer or A/CA and which are themselves subject to copyright or similar restrictions.

2.41.2   As long as the Contractor has been properly paid for his work under this contract the Employer is free to use the CDD for any purpose relating to the Works. For example, where the Contractor's employment has been validly terminated the Employer may use the CDD to allow the remaining construction works to be completed. Also the Employer may use the CDD for the maintenance of the Works on completion, repair works, promotional literature, etc.

If the Employer subsequently has an extension added to the Works, the CDD may be used to enable the new works to be designed and executed, e.g. to locate existing services, load-bearing members, details of the structure where openings are proposed, etc. However, the Employer may not use the CDD for the actual extension work. For instance, it would not be permissible for the Employer to copy and replicate a Contractor Designed lighting system – without gaining the Contractor's consent.

2.41.3   The Contractor's liability in relation to the CDD is limited solely to the purpose for which they were prepared. For example, the Contractor would not be liable if a goods lift with a required capacity of 2 tonnes failed when grossly overloaded, nor would the Contractor be liable for any failure if the Employer used a design on another project without the Contractor's consent.

Section 3: Control of the Works

Access and Representatives
Access for Architect/Contract Administrator

3.1   Under clause 2.4 the Contractor is given possession of the site, which means that, technically, the A/CA must have the Contractor's permission before he is able to come onto the site. The A/CA obtains permission through clause 3.1, which requires that he (and persons authorised by the A/CA) be given reasonable access to the Works. The A/CA's right of access is also to be extended to include the Contractor's, and sub-contractors', workshops and premises where work is being prepared for this contract. However, access to workshops and premises may be limited, if necessary, to protect a Contractor's or sub-contractor's proprietary rights (e.g. a patented process).

Person-In-Charge

3.2   The Contractor is required to have on site an individual defined as a 'competent person-in-charge'. This person should be on site at all times during normal working hours, and any A/CA's instructions or clerk of works directions received by this person are deemed to have been issued to the Contractor (i.e. see clause 1.7; in this instance the instruction would have been given 'by any effective means').

Employer's Representative

3.3   For administrative purposes the Employer may appoint an individual from within his organisation (or possibly an external appointment) to act on his behalf, i.e. an Employer's representative. The Employer must give a written notice to the Contractor identifying the Employer's representative and the date when his role becomes operative. From this date the Employer's representative will act as the Employer. If the Employer has given his representative limited powers under the contract then he must clearly inform the Contractor, in the notice, of the areas where the representative is not empowered to act. It is advised that the Employer should not consider appointing the A/CA or QS as his representative as it may cause some confusion between the roles. For example, if the A/CA were to be appointed as the Employer's representative it would be difficult for the Contractor to appreciate when the A/CA was acting in his normal capacity and when he was acting as the Employer's representative.

The Employer may remove or replace the Employer's representative at any time, but must inform the Contractor through a written notice. The Contractor has no right of objection to any replacement chosen by the Employer.

Clerk of Works

3.4   The Employer may appoint a clerk of works who will work under the direction of the A/CA. The role of the clerk of works is to inspect the Works and report back to the Employer or A/CA, but he has no authority to approve the Works. The Contractor should act reasonably so as to allow the clerk of works to carry out his inspection. Although the clerk of works may issue directions to the Contractor, they are not valid unless confirmed by the A/CA within 2 working days. A clerk of works may only issue a direction on matters within the A/CA's authority (i.e. he cannot exceed the A/CA's authority).

Replacement of Architect/Contract Administrator or Quantity Surveyor

3.5.1 If, for any reason, the A/CA or Quantity Surveyor named in the Articles ceases to be employed on the project, then the Employer is to nominate a replacement. The Employer is to give the Contractor a notice of the replacement within 21 days from when either the A/CA or Quantity Surveyor ceased to be employed on the project. The Contractor has the right to issue a counter-notice objecting to this replacement. The counter-notice must be issued within 7 days of the nomination notice. If the Employer accepts the reasons given by the Contractor, he must find an acceptable replacement. If the Employer does not accept a Contractor's objection, an adjudicator may resolve the dispute in accordance with Article 7.

The Contractor has no right of objection where the Employer is a Local Authority and the nominated replacement is an official of the Local Authority.

(See Croudace Ltd v The London Borough of Lambeth, 1986)

3.5.2 A replacement A/CA cannot ignore or override actions taken by the previous A/CA unless the contract conditions would have allowed the previous A/CA to do this if he were still in post. If, for example, the previous A/CA had issued a Certificate of Making Good to a Section of work and it was subsequently discovered that some defects had not been rectified, it is not possible for the new A/CA to revoke that certificate. However, if the previous A/CA had included the value of defective work in a previous Interim Certificate, the new A/CA may remedy that by deducting the value of defective work from the next Interim Certificate. The reason being that the contract conditions allow for errors in Interim Certificates to be subsequently rectified.

Contractor's Responsibility

3.6 The purpose of this clause is to reinforce the principle that it is the Contractor who is responsible for ensuring that the Works are carried out and completed in accordance with the Contract Documents. The fact that a clerk of works may have observed the work being carried out, or that the A/CA has inspected the work on site or at workshops or premises, or that the work has been paid for in Interim Certificates, does nothing to diminish the Contractor's responsibility.

It is important to note that a possible exception to the above responsibility could arise where the work has been specified in the Contract Documents as having to be carried out to the A/CA's reasonable satisfaction (see clauses 1.10.1.1 and 2.3.3). If work has been specified as being to the approval of the A/CA, then failure by the A/CA to state his dissatisfaction within a reasonable time would imply that the work had been accepted (see clause 3.20).

Sub-Letting
Consent to Sub-Letting

3.7.1 Although the Contractor may sub-let the Works or portions of the Works to sub-contractors, he is still responsible to the Employer for the standard of work and the giving of statutory notices as detailed in clause 2.1. The Contractor must obtain the A/CA's written consent before sub-letting any work. The A/CA may not unreasonably withhold or delay his consent.

3.7.2 The Contractor may sub-let the design work relating to a CDP where he has obtained the written consent of the Employer. The Employer is not to unreasonably withhold or delay his consent. Although the design may have been sub-let with the Employer's permission, this does not remove the Contractor's obligations to the Employer with regards the design work. Particular attention is drawn to the Contractor's obligations under clauses 2.2 and 2.19, but they are not exclusive and other obligations may arise under other parts of the contract.

3.7.3 This clause explains that, where Statutory Undertakers are working on site in their statutory capacity (e.g. the provision of a power cable and meter to a property), they are not sub-contractors for the purposes of this contract and the procedures set out in clauses 3.7 to 3.9 cannot be applied to their work.

List in Contract Bills

3.8.1 On occasions the bill of quantities may require that certain work must only be carried out by one of a number of firms that are named in a list contained within the bills or attached to the bills; it is the Contractor's responsibility to price the items in the bills and select which of the named firms is to execute the work.

3.8.2 There must be at least three names on the list and, at any time prior to the execution of a binding sub-contract, further names may be added by the Employer (or A/CA on his behalf) and Contractor. If further names are to be added, the consent of the other party must be obtained, and this consent cannot be unreasonably withheld or delayed. Footnote [39] advises that the consent should be confirmed in writing.

3.8.3 If, prior to the execution of a binding sub-contract agreement, the number of firms prepared to carry out the work drops to less than three, the Employer and Contractor may agree the addition of further names as necessary. The agreement should be in writing and should not be unreasonably withheld or delayed. Alternatively, the naming procedure may be allowed to lapse and the Contractor may carry out the work himself or sub-let it to a sub-contractor of his choice as long as he obtains the necessary consent under clause 3.7.

3.8.4 Any person appointed from a list under this naming procedure is a sub-contractor, i.e. they have no special status that differentiates them from the Contractor's normal sub-contractors.

Conditions of Sub-Letting

3.9 When a Contractor employs a sub-contractor it is a requirement that certain terms must be incorporated into the sub-contract agreement. If the Contractor uses the JCT standard form of domestic sub-contract SBCSub or SBCSub/D, then this requirement is automatically fulfilled. If a Contractor uses a non-standard form of sub-contract, then it must incorporate the terms set out in clauses 3.9.1 and 3.9.2, as detailed below:

3.9.1 a term must be incorporated into the sub-contract which states that a sub-contractor's employment will be terminated immediately upon the termination of the Contractor's employment under this contract (see SBCSub/D clause 7.9);

3.9.2 the following conditions must also be incorporated:

    3.9.2.1 where the sub-contractor has unfixed materials on site, for incorporation into the Works, the materials may not be removed from site without the Contractor's consent. Consent shall not be unreasonably delayed or withheld (see SBCSub/D clause 2.15.1);

        3.9.2.1.1 where the sub-contractor's unfixed materials have been included in an Interim Certificate and the Employer has paid the amount *properly* due in the certificate, then the sub-contractor materials become the property of the Employer. The sub-contractor shall not deny that such materials have become the property of the Employer.

          This clause is an attempt to overcome the problems raised by the case of Dawber Williamson v Humberside C.C. There is some doubt as to the effectiveness of this clause because of the principle of 'privity', i.e. the Employer cannot enforce a benefit in the sub-contract and the sub-contractor cannot be bound by an obligation in the main contract. This problem could be overcome if the JCT were to consider using the Contracts (Rights of Third Parties) Act to give the Employer an enforceable benefit in the sub-contract. If the Employer does not pay the amount properly due in the Interim Certificate (i.e. withholds a sum of money without issuing the appropriate withholding notice) then any sub-contract materials included in the certificate will not become the Employer's property (see SBCSub/D clause 2.15.2);

        3.9.2.1.2 if the Contractor pays the sub-contractor for sub-contractor materials before they have been paid for by the Employer through an Interim Certificate, the materials become the property of the Contractor (see SBCSub/D clause 2.15.3);

          To appreciate the problems an Employer may face when paying for a sub-contractor's unfixed materials see Dawber Williamson Roofing Ltd v Humberside County Council (1980).

3.9.2.2 the sub-contractor is to allow the A/CA and his representatives access to his workshops and premises where work is being prepared for this contract. This term must be incorporated into the sub-contract conditions in order that the Contractor may comply with his obligations under clause 3.1 (see SBCSub/D clause 3.9);

3.9.2.3 if the Contractor fails to pay the sub-contractor monies that are properly due, then the sub-contractor is entitled to receive interest on the overdue amount at 5% above the Bank Base Rate (see clause 1.1 Definitions; Interest Rate) current at the date the payment became overdue. This provision to allow interest on overdue payments does not remove a sub-contractor's right of suspension or termination because of the Contractor's failure to comply with the payment provisions (see SBCSub/D clause 4.10.5);

3.9.2.4 where there is a requirement for a sub-contractor to provide a collateral warranty as set out in the main contract (Part 2: Third Party Rights and Collateral Warranties) the warranty must be executed and delivered by the sub-contractor within 14 days of receiving the Contractor's written request (see SBCSub/D clause 2.26.1);

3.9.2.5 clauses 3.9.2.1.1 and 3.9.2.1.2 do not override or modify the procedures for property in materials passing to the Contractor under clause 4.17.2.1: off-Site Materials (see SBCSub/D clause 2.15.4).

Architect/Contract Administrator's Instructions
Compliance with Instructions

3.10 The Contractor has to comply with all A/CA's instructions (with three exceptions as explained in the following sub-clauses), as long as the instruction is valid. An A/CA may only issue an instruction where the contract conditions allow him to do so. If an A/CA attempts to exceed that authority, then the instruction is invalid. There is no set period of time in which a Contractor is to comply with an instruction, although there is a requirement for the Contractor to comply 'forthwith'. The dictionary definition of 'forthwith' is 'immediately' but because of the nature of construction work, this needs to be translated into requiring the Contractor to act *reasonably soon,* bearing in mind the current progress of the Works and the nature of the instruction.

3.10.1 This clause modifies the Contractor's obligation to comply with all A/CA's instructions. Where the A/CA issues a variation instruction under clause 5.1.2 (this relates to obligations and restrictions regarding the site, working hours and working space, etc.) the Contractor does not have to comply as long as he submits, to the A/CA, a reasonable objection in writing. Note that this right of reasonable objection applies only to a clause 5.1.2 variation and not to A/CA's instructions in general.

3.10.2 Similarly, a Contractor is not to execute work identified in a Schedule 2 instruction until he receives a 'Confirmed Acceptance' from the A/CA or a subsequent instruction for the work to be carried out under clause

5.3.2. A Schedule 2 instruction is another instance where the Contractor need not comply, see clause 5.3.1.

3.10.3 If a Contractor receives a direction (under clause 2.2.2) or an instruction which he considers will have an adverse effect on the design of the CDP, he need not comply with it immediately. Within 7 days of receiving the direction or instruction he should provide the A/CA with a written notice identifying how the direction or instruction would have an adverse effect on the design. As a result, the Contractor need not comply with the direction or instruction unless the A/CA confirms that the Contractor is to comply.

Non-compliance with Instructions

3.11 This clause is commonly misinterpreted. If the A/CA wishes the Contractor to comply with an instruction that has already been issued, he must follow up the instruction with a written notice in which he requests the Contractor to comply with the instruction identified in that notice. If the Contractor fails to proceed with an instruction *within 7 days from receipt of the 'follow up' written notice*, then the Employer may engage outside contractors to carry out the work, etc. and may recover the extra cost from the Contractor. The cost is to be recovered by making a deduction to the Contract Sum, e.g. the cost may be recovered from monies due in the next Interim Certificate (clause 4.4). The following example illustrates how this procedure should operate: an instruction is issued on 1st November, on 12th November the A/CA becomes aware that the instruction has not been carried out and issues a written notice to the Contractor asking him to comply with the instruction. On 19th November the Contractor has still not carried out the instruction, on 20th November the Employer *may* employ outside contractors to carry out the work.

Instructions To Be in Writing

3.12.1 This sub-clause sets out a basic requirement that all instructions issued by the A/CA are to be in writing which is obviously good practice as it provides concrete evidence of what the A/CA required.

3.12.2 If the A/CA issues a verbal instruction it will have no immediate effect. However, on receiving a verbal instruction the Contractor should, within 7 days of its receipt, write to the A/CA confirming the details of the instruction. In return the A/CA may within 7 days from receipt of the confirmation write back to the Contractor stating he does not accept the confirmation notice, in which case no instruction will exist. If the Contractor receives no written notice of dissent from the A/CA within the second 7-day period then a valid instruction exists based upon the Contractor's written confirmation. The instruction will officially exist from the end of the second 7-day period of notice.

3.12.3 This sub-clause covers the situation where the A/CA issues a verbal instruction and subsequently confirms it in writing. If the A/CA provides his own written confirmation within 7 days of having issued a verbal instruction, the Contractor's obligation to send a written confirmation is removed. The instruction will be valid from the date of the A/CA's written confirmation.

3.12.4 If the Contractor complies with a verbal instruction but neither the Contractor nor A/CA confirmed the instruction in writing, the A/CA is allowed to issue a retrospective confirmation. This retrospective action may take place at any time up to the issue of the Final Certificate, whereupon the A/CA's instruction will be deemed to have been effective on the date of issue of the verbal instruction.

Provisions Empowering Instructions

3.13 If the Contractor receives an instruction (either written or verbal) from the A/CA which he thinks may be invalid, then he may query it. The Contractor may ask the A/CA to provide a written notice identifying the contract condition that allows the A/CA to issue such an instruction. The A/CA is to respond immediately. It would appear that, if the A/CA responds giving an incorrect or inappropriate clause reference but the Contractor nevertheless complies with the instruction, then, for the purposes of this contract, the instruction will be deemed valid. However, if the Contractor is unhappy with the A/CA's response, he may refer the issue to adjudication.

Instructions Requiring Variations

One of the main reasons why an A/CA may issue an instruction is to change some part of the original design, i.e. he issues a variation. This clause provides some basic administrative detail in relation to variations.

3.14.1 This sub-clause gives the A/CA the authority to issue an instruction that would result in a variation (see clause 5.1 for a Definition of Variations).

3.14.2 This sub-clause confirms that the Contractor has the right to make a reasonable objection (also see clause 3.10.1) to any instruction issued under clause 5.1.2, i.e. a variation instruction changing an imposition or restriction imposed by the Employer.

3.14.3 Where the A/CA wishes to vary an item of CDP work, then the variation instruction will result in an alteration to the Employer's Requirements. The purpose of this clause is to clearly show that the Contractor is entitled to receive payment (or deduction) for the change to the CDP work.

3.14.4 If the Contractor has varied the work and there is no A/CA's instruction (e.g. he may have complied with a clerk of works direction which was never properly confirmed) the A/CA may, if he wishes, issue a retrospective instruction to cover that work.

3.14.5 A variation issued or subsequently sanctioned by the A/CA will not vitiate the contract. That is, it will not invalidate the agreement between the Employer and Contractor: although the Contractor has agreed to carry out the Works in accordance with the Contract Documents, he has also agreed that the requirements and detail within the documents may be varied.

Postponement of Work

3.15   The A/CA may issue an instruction requiring the Contractor to postpone carrying out any of the Works. This does not omit the work, but merely delays the execution of the work identified in the instruction. At first glance it may appear that clause 3.15 is duplicating the procedures contained within clause 2.5 (i.e. deferred possession). However, clause 3.15 allows the A/CA to only instruct the Contractor to stop or delay working on a Section of the Works whereas clause 2.15 allows the Employer to delay the Contractor from taking possession of the site or parts of the site.

Instructions on Provisional Sums

3.16   This clause explains how Provisional Sums should be dealt with under the contract conditions. Whenever a Provisional Sum exists, in the Contract Bills or Employer's Requirements, the A/CA is required to issue the Contractor with an instruction detailing what is to be done with the Provisional Sum. The instruction may inform the Contractor of the work to be executed against the Provisional Sum, or it may state that the Provisional Sum is to be omitted and no work is to be carried out.

Inspection – Tests

3.17   This clause allows the A/CA to instruct that work may be opened up for inspection or that work or materials may be tested. If the test and/or inspection is satisfactory, the cost will be added to the Contract Sum unless already allowed for in the Contract Bills, e.g. the SMM7 provides an item for testing certain elements of the work, such as plumbing and drainage, and the Contractor has the opportunity of pricing the cost of testing at tender stage. If the test and/or inspection proves unsatisfactory, the Contractor will bear the costs of opening up the work, carrying out the test and making good.

Work Not in Accordance with the Contract

3.18   This is a complicated clause and requires careful reading to understand all the relevant procedures. The purpose of this clause is to identify what actions the A/CA and Employer may take once an item of defective work has been discovered. The available options are identified below:

    3.18.1   On discovering an item of defective work, the most obvious solution would be to issue an instruction for it to be removed, and for the Contractor to re-execute the work. That is the procedure allowed for in this clause.

    3.18.2   Alternatively, there may be occasions where an item is technically defective (i.e. it does not comply with the specification) but it is still adequate for its purpose. In this case the A/CA may give the Contractor a written notice that some, or all, of the defective work may remain and an appropriate deduction be made to the Contract Sum. If the A/CA is to adopt this approach he must first discuss the implications with the

Contractor (although he does not require the Contractor's consent) and obtain the Employer's consent. The A/CA cannot use this procedure for defective work which is part of the Contractor's CDP work.

3.18.3   Where there has been an instruction issued for the removal of defective work, or an instruction issued for defective work to remain, it may be necessary on occasions to alter a part of the Works or design as a direct result of the instruction. As a consequence, the A/CA may issue a variation instructions. For example, the Contractor may have fixed plasterboarding of a lesser thickness than that specified, and as a consequence the work does not comply with the fire regulations. The A/CA instructs that the defective boarding is to remain but issues a variation instructions for the Contractor to apply two coats of plaster in lieu of one, so as to achieve the necessary fire rating. Before the A/CA issues this consequential variation it is a requirement that he consults with the Contractor, although he does not need the Contractor's consent. The Contractor will receive no payment for this variation nor is he entitled to any extension of time or claim for loss and/or expense associated with the variation work.

3.18.4   If an item of defective work has been discovered and there is a possibility that a similar defect may be present in other areas of the work, the A/CA may issue an instruction to test further areas of work to discover the extent of the defect. Any instruction issued by the A/CA for further testing must be 'reasonable in all the circumstances'. It may at times be difficult to decide what is reasonable, therefore the JCT has produced a Code of Practice (Schedule 4, attached to the back of the conditions) to try and help the A/CA, in consultation with the Contractor, to determine what is a reasonable level of testing. The A/CA will issue the instruction under clause 3.17, but the normal payment terms are modified by clause 3.18.4, to the effect that, regardless of the outcome of the tests, the Contractor will have to pay all costs (as long as the instruction given was reasonable) and will not be entitled to any claim for loss and/or expense associated with the test. However, if the tests prove the work, etc. to be satisfactory the Contractor is entitled to request an extension of time if the additional testing has delayed the progress of the Works.

Workmanship Not in Accordance with the Contract

This clause deals with the *manner* in which the Contractor carries out the Works and not the actual workmanship. The reference to 'workmanship' in the clause heading is misleading – in previous editions of the JCT the heading used for this clause referred to 'workmanlike manner'. Workmanlike manner is concerned with how the Contractor is carrying out the work, e.g. is the work properly scaffolded, is plant being used with the correct screen and guards, is plant being used in a safe and correct manner?

3.19   If the Contractor fails to comply with his obligation to carry out the Works in a proper and workmanlike manner or in accordance with the Construction Phase Plan, the A/CA may (after consulting with the Contractor) issue an instruction to remedy this. As long as the instruction is reasonable, the Contractor is not entitled to any additional money as a result of the instruction, nor is he entitled to an extension of time or payment for loss and/or expense.

(See Greater Nottingham Co-operative Society Ltd v Cementation and Others, 1988)

Executed Work

3.20   This clause is frequently misinterpreted. It is only relevant where an item of work or material, etc. has been specified in the Contract Documents (e.g. in the preambles bill) as being 'to the Architect's/Contract Administrator's approval'. In this event the problem for the Contractor is that he will not know whether the work or material is satisfactory unless the A/CA informs him. Where, therefore, an item has been described in the Contract Documents as having to be 'to the Architect's/Contract Administrator's approval' the A/CA must express any dissatisfaction with the standard of work or materials within a reasonable time, and explain why he is dissatisfied. A question still remains as to what may be considered a reasonable time. If no agreement can be reached between the parties, as to the timing of the notification, the dispute may be referred to adjudication, etc. If the A/CA does not express dissatisfaction within a reasonable time then it is deemed that he must be satisfied with the work.

Note: this clause will have no relevance where workmanship, materials or goods have been fully described in the Contract Documents or where no standard has been specified.

3.21   The A/CA may not instruct the Contractor to terminate the employment of any operative on site, but if he is unhappy with the performance, etc. of an operative, he may instruct the Contractor that the operative is to be excluded from the site. Any such instruction must be reasonable.

Antiquities
Effect of Find of Antiquities

3.22   Fossils, antiquities, objects of interest or value, etc. found on the site during the progress of the Works belong to the Employer. When the Contractor discovers such an item he must:

3.22.1   try and ensure the object is not disturbed and stop work if necessary to preserve the object or to allow for its excavation and removal;

3.22.2   take all necessary steps to allow the object to remain in its discovered position and condition;

3.22.3   inform the A/CA or clerk of works of the discovery, giving details of its exact location.

Instructions on Antiquities

3.23   Where the Contractor reports the discovery of an object, the A/CA may issue an instruction allowing a third party to examine, excavate and remove the object.

Loss and Expense Arising

3.24   If the A/CA considers that the Contractor has incurred direct loss and/or expense as a result of complying with clause 3.22 or an instruction issued under clause 3.23, the A/CA is to evaluate the claim (or the QS may be instructed to assess the claim). As and when any direct loss and/or expense is determined, it should be added to the Contract Sum, i.e. payable in the next Interim Certificate. There is a proviso that this loss and/or expense procedure is only operated where the loss and/or expense incurred would not be reimbursable under any other contract condition.

CDM Regulations
Undertakings to Comply

3.25   This clause identifies that both parties (Employer and Contractor) are deemed to be aware of the CDM Regulations and, furthermore, that each party will comply with the CDM Regulations as far as they relate to the Works and the site. Where the project is notifiable, the Employer and Contractor are bound to comply totally with the following obligations:

(*Note*: a project is notifiable where the Construction Phase is likely to take longer than 30 days to complete, or it is likely to require more than 500 person days to complete the construction work. Where a project is notifiable, it will have to comply with Part 3 of the CDM Regulations.)

3.25.1   the Employer is responsible for ensuring that the CDM Co-ordinator carries out all his relevant duties. On a project where the Contractor is not the 'Principal Contractor' (see Article 6) the Employer is responsible for ensuring that the Principal Contractor carries out all his relevant duties;

3.25.2   where the Contractor is the Principal Contractor, and for as long as he remains the Principal Contractor, he must comply with the following:

(*Note*: full details of the Principal Contractor's obligations can be found in paragraphs 21 to 24 of the CDM Regulations.)

3.25.2.1   A Construction Phase Plan is to be prepared and given to the Employer before any construction work may be started. If the Construction Phase Plan is subsequently altered, the Employer and CDM Co-ordinator must be informed of the alteration. Normally the CDM Co-ordinator will be the A/CA, but where this is not the case the Contractor must also inform the A/CA of any alterations.

3.25.2.2   Welfare facilities in compliance with Schedule 2 of the CDM Regulations must be provided. These facilities are to be provided at the start of the construction work and maintained throughout until the end of the construction phase.

(*Note*: Schedule 2 of the CDM Regulations lists the facilities required to be on site and the standard of the facilities to be provided. The specific facilities that are identified are as follows: sanitary conveniences, washing facilities, drinking water, changing rooms and lockers, rest facilities.)

Footnote [41] identifies that there is an obligation on Contractors to comply with Schedule 2 of the CDM Regulations as far as it is reasonably practicable, even where the project is not notifiable or the Contractor is not the Principal Contractor.

3.25.3 In a situation where the Contractor is not the Principal Contractor, he must promptly inform the Principal Contractor of the identity of every sub-contractor he engages to carry out work on this project. Similarly, where a sub-contractor sub-lets a portion of the sub-contract works, the Contractor is to be notified so that he in turn can notify the Principal Contractor of the identity of the sub-sub-contractor.

3.25.4 Where the Contractor is the Principal Contractor, he is obliged to provide certain information to the CDM Co-ordinator to enable the preparation of the health and safety file. The Contractor is to promptly provide such information as is reasonably required and requested (in writing) by the CDM Co-ordinator. The Contractor is also responsible for obtaining the necessary information from his sub-contractors and passing it on to the CDM Co-ordinator. Where the Contractor is not the Principal Contractor, he is still under the same obligation to provide the necessary information, but in this instance the information is to be passed to the CDM Co-ordinator via the Principal Contractor.

Appointment of Successors

3.26 Under Articles 6 and 7 it is permissible for the Employer to appoint a replacement CDM Co-ordinator and/or Principal Contractor. If the Employer exercises this option he must immediately give the Contractor a written notification identifying the name and address of the new appointee. Where the Contractor was originally specified as the Principal Contractor but has been subsequently replaced, he must now comply with all reasonable requirements of the replacement Principal Contractor. In complying with these requirements the Contractor cannot claim any extra payments or extensions of time.

Section 4: Payment

Contract Sum Adjustments
Work Included in Contract Sum

4.1 The Contract Sum is based on the specification and quantities set out in the Contract Bills, therefore if the drawings, schedules, instructions, etc. require the Contractor to provide something different from what is specified in the Contract Bills, the Employer will have to pay for the difference. Where the project includes a Contractor's Designed Portion, the quality and quantity of that work will be set out in the CDP Documents.

Adjustment Only Under Conditions

4.2 The Contract Sum may only be adjusted in accordance with an express term in the contract, therefore it is advisable that the QS and Contractor are fully aware of all the contract terms which allow an adjustment to be made to the Contract Sum. If there is an error of description or quantity in the Contract Bills, then that error must be rectified under clause 2.14. However, if there is an arithmetic error (e.g. this could be an error by the Contractor when extending unit rates or totalling pages or transferring totals to summary pages) and it has not been spotted at the pre-contract stage, then it is deemed that both the Employer and Contractor have accepted the error – it will not now be corrected.

(See Henry Boot Construction Ltd v Alstom Combined Cycles Ltd, 2000)

Items Included in Adjustments

4.3.1 This clause describes how the Contract Sum is to be adjusted. Initially the Contract Sum will be adjusted to take account of the following:

4.3.1.1 the value of variation work carried out under clause 5.2.1 where the value has been agreed between the Employer and Contractor;

4.3.1.2 the value of work carried out under an accepted Schedule 2 Quotation plus the value of any associated variations, i.e. instructions that have been issued under clause 5.3.3; and

4.3.1.3 any change in cost to the Contractor of the premium for Terrorism Cover, all in accordance with Schedule 3 paragraph A.5.1. This clause is only relevant where the Contractor is responsible for the Works insurance through Insurance Option A.

4.3.2 Subsequently, the following items are identified that are to be deducted from the Contract Sum:

4.3.2.1 all Provisional Sums and the value of any Approximate Quantities that may have been included in the Contract Bills. This will also apply to the Employer's Requirements where the Contractor is carrying out CDP work;

4.3.2.2 the value of any work that has been omitted through the issue of a variation (i.e. clauses 5.6.2 and 5.8.3). Also, if under the operation of clause 5.9 an item of 'other' work is to be revalued in accordance with Section 5, then the original bill rate for that item is to be omitted;

4.3.2.3 any amount to be deducted or payable to the Employer as a result of the operation of any of the following clauses:

clause 2.10, inaccurate setting out of the Works by the Contractor;

clause 2.38, where the A/CA instructs that defective work identified during the Rectification Period is not to be remedied;

clause 3.11, where the Contractor has failed to comply with an instruction of the A/CA, and the Employer has engaged other contractors;

clause 3.18.2, where the A/CA instructs that defective work discovered before practical completion is not to be remedied;

clause 6.15.2, failure by the Contractor to comply with Joint Fire Code remedial measures.

Fluctuations Options A, B and C, i.e. fluctuation credits payable to the Employer.

4.3.2.4 finally, any other sum which the contract requires to be deducted from the Contract Sum – a catch-all phrase to mop up any item not included in the above list.

4.3.3 The following list identifies amounts that are to be added to the Contract Sum:

4.3.3.1 payments due to the Contractor under the following clauses:

clause 2.21, payment for royalties, payment of statutory fees, etc.;

clause 2.23, payment for patent rights;

clause 3.17, payment for tests, inspections and making good;

clause, 6.5.3, payment of premium for insurance required by the Employer under clause 6.5.1;

4.3.3.2 variations valued by the QS under Section 5, including 'other' work, which has been revalued under clause 5.9. Variations of omissions are obviously excluded from this section;

4.3.3.3 the value of work executed or costs incurred by the Contractor following an A/CA's instruction as to the expenditure of a Provisional Sum included in the Contract Bills (or Employer's Requirements where relevant). The value of any Approximate Quantities included in the Contract Bills (or Employer's Requirements where relevant);

4.3.3.4 any amounts claimed under clause 3.24 (loss and/or expense in connection with antiquities) and 4.23 (loss and/or expense in connection with a disruption to the regular progress of the Works);

4.3.3.5 cost of insurance premiums paid by Contractor where the Employer has failed to take out the insurance (see Insurance Option B.2.2 and Option C.3.3), and any additional premium payable by the Contractor as a result of the Employer exercising his right of 'Early Use' (see clause 2.6.2);

4.3.3.6 amounts payable to the Contractor under Fluctuations Options A, B or C;

4.3.3.7 any other amount that is required to be added to the Contract Sum.

Taking Adjustments into Account

4.4 At times there will be a statement in the contract conditions to the effect that a deduction or addition should be made to the Contract Sum (e.g. clauses 2.10, 2.21 and 3.18.2). Clause 4.4 sets out the timing of these payments. As soon as the required deduction or addition has been assessed, it is to be taken into account in the next Interim Certificate; it may be a partial adjustment on account or the total adjustment, depending upon the circumstances. The important point is that, when reference is made to making an addition to, or deduction from, the Contract Sum, this calculation does not have to wait until the final account is being prepared but should be acted upon promptly.

Final Adjustment

4.5.1 Not later than 6 months from the date of practical completion (or the last Section Completion Certificate) the Contractor is to provide the A/CA with all the necessary documentation to allow the final account to be prepared. Alternatively, the A/CA may instruct the Contractor to send the documentation to the QS, who will normally prepare the final account.

4.5.2 Once the A/CA (or QS) has received the above documents, then within 3 months the following should take place:

4.5.2.1 the A/CA (or QS) is to calculate any loss and/or expense claimed by the Contractor under clause 3.24 or 4.23, unless the amounts have already been calculated, and

4.5.2.2 the QS is to prepare a statement of all the other adjustments which are to be made to the Contract Sum, all in accordance with clause 4.3 (i.e. the final account excluding the loss and/or expense claims under clause 3.24, etc.).

As soon as either of the above the documentation is completed, the A/CA is to send a copy to the Contractor.

Certificates and Payments
VAT

4.6.1 The Contract Sum is exclusive of VAT; therefore, whenever financial adjustments are made to the Contract Sum (e.g. variations) VAT is ignored. However, that does not mean that VAT can be completely ignored; when making payments to the Contractor, the Employer must include the value of any necessary VAT payments. This is a matter that the Employer and Contractor will have to sort out between themselves to ensure that VAT Regulations are complied with and the correct VAT is charged and collected.

4.6.2 Under VAT Regulations a Contractor is unable to claim his input tax from the Customs and Excise for the supply of goods or services that are exempt from VAT. In such a situation the VAT charged to the Contractor by a supplier or sub-contractor would not be recoverable under the VAT Regulations and would have to be recovered from the Employer as a cost built into the unit rate. Consequently, if a supply of goods or services is exempt as a result of a change in the law after the Base Date, the Employer is to pay the Contractor an amount equivalent to the input tax paid by the Contractor when he purchased the goods or services.

Construction Industry Scheme (CIS)

The Construction Industry Scheme sets out the legislation dealing with payment and associated taxation payable by sub-contractors working in the construction industry. The legislation was introduced to try and reduce the high level of tax evasion that was occurring in the construction industry.

4.7 Where an Employer carries out construction work in his own right, he may be defined as a 'contractor' in accordance with the Construction Industry Scheme. Where this is the case, the Employer's payments to the Contractor must be made in accordance with the CIS (i.e. the Contractor will be deemed a sub-contractor of the Employer for the purposes of the CIS and will have to provide the necessary documentation to allow the Employer to pay monies in full without making deductions for tax). For the purposes of this clause it is important that the Fourth Recital and Contract Particulars are properly completed.

Advance Payment

4.8 This is an optional clause which makes provision for the Employer to make an advance payment to the Contractor. The Contract Particulars provide details of the payment to be made, how it is to be paid off by the Contractor and whether the Employer requires the protection of a bond. If the Contract Particulars require the Contractor to take out an advance payment bond then the advance payment is conditional upon the Contractor providing a bond in accordance with the procedures set out in Schedule 6, Part 1. The bond must be provided by a surety (i.e. a bank) that meets the Employer's approval.

It is important to note that this clause does not apply where the Employer is a local authority – see Contract Particulars.

Issue of Interim Certificates

4.9.1 The A/CA is responsible for issuing Interim Certificates in accordance with the time schedule set out in the Contract Particulars. The certificate is to state the amount due to the Contractor from the Employer. Also the A/CA is to provide details about what has been included for payment and how the amount was calculated. These details will normally be provided to the A/CA by the QS in his interim valuation.

4.9.2 Until practical completion has been achieved Interim Certificates must be issued in accordance with the date(s) provided in the Contract Particulars. After practical completion, Interim Certificates are to be issued as and when necessary (i.e. when further monies become due to the Contractor) although the A/CA may not be required to issue an Interim Certificate before a full calendar month has elapsed since the previous certificate. The A/CA must issue a certificate after the end of the Rectification Period or after the issue of the Certificate of Making Good, whichever is the later, to allow the final release of Retention to the Contractor.

Amounts due in Interim Certificates

4.10 This clause identifies the format and the detail of the Interim Certificate. Unless the Contractor and Employer have agreed a system based upon stage payments (i.e. a stage of work to be paid for only when it has been completed) the Gross Valuation is to be calculated in accordance with clause 4.16. The Interim Certificate is to show the gross value of work as set out in clause 4.16. Certain sums may be deducted from the gross value, i.e.:

4.10.1 Retention may be deducted in accordance with the rules laid down in clauses 4.18 to 4.20;

4.10.2 where an advance payment(s) has been agreed, the cumulative repayments are to be deducted from the gross value as and when they fall due (dates will be stated in the Contract Particulars);

4.10.3 finally, the total amount that had been stated as being due in previous Interim Certificates is to be deducted.

The following example illustrates the operation of clause 4.10:

Interim Certificate no. 7	£	£
Gross valuation		1,400,000.00
Deduct		
Retention	42,000.00	
Advance payment	50,000.00	
Previous interim payments	1,114,000.00	
Sub-total	1,206,000.00	1,206,000.00
Net valuation		**194,000.00**

Interim Valuations

4.11 It is the Quantity Surveyor's responsibility to prepare interim valuations. Although the decision as to whether or not an interim valuation is required rests with the A/CA the conditions of engagement between the QS and the Employer would normally require the QS to prepare regular interim valuations.

Where the contract is operating under Fluctuation Option C it is mandatory that the QS prepares an interim valuation prior to the issue of the Interim Certificate because a detailed interim valuation is required to allow for the proper assessment of fluctuations under Option C (Schedule 7, paragraph C.2). Although it is not specified, a QS would also have to prepare an interim valuation on receipt of a Contractor's application (see clause 4.12).

Application by Contractor

4.12 The Contractor is entitled to submit an application for payment for the value of work completed. The application should be the Contractor's assessment of the Gross Valuation and as a consequence it should be calculated in accordance with clause 4.16. The application must be submitted, to the QS, no later than 7 days before the date of issue of the Interim Certificate. On receipt of the Contractor's application the QS is required to prepare an interim valuation. If the QS disagrees with any aspect of the application he must provide the Contractor with a statement giving details of the disagreement. The statement (to be prepared at the same time as the interim valuation) is to be prepared in the same amount of detail as the Contractor's application and it should identify where the disagreement has occurred.

Interim Certificates – Payment

4.13.1 The final date by which the Employer should pay the Contractor the amount identified in an Interim Certificate is 14 days from the date of issue of that certificate.

4.13.2 Where, under the terms of the contract, the Employer is entitled to recover monies from the Contractor, he may do so by withholding/deducting monies from the Interim Certificate. Where Retention monies are released to a Contractor through an Interim Certificate, the Employer has the right to withhold/deduct monies from the Retention, despite the fact that the Employer has been acting as a trustee and has been holding the Retention in trust for the Contractor. The argument used to allow the Employer to access the Retention money is that, once the Retention monies are released to the Contractor through an Interim Certificate (see clause 4.20), they lose their trust status and the Employer may now make a claim against these monies due to the Contractor.

4.13.3 To comply with the HGCRA the Employer is required to confirm the amount to be paid to the Contractor, what the payments are for and how they have been calculated. The Employer is required to do this within 5 days from date of issue of the Interim Certificate. Unless the Employer wants to amend the amount due under the A/CA's Interim Certificate, he is likely to ignore this requirement (see JCT 80 Amendment 18 for a further explanation of this) although this is technically a breach of contract. Where the Employer fails to provide the notice required by this clause the Contractor is entitled to the monies stated, as due, in the Interim Certificate (see clause 4.13.5).

4.13.4 If the Employer intends to pay a lesser amount than that stated in the Interim Certificate (or the 4.13.3 notice if given) he must notify the Contractor at least 5 days before the final date for payment. In the written notice the Employer must specify the amount to be withheld/deducted, the ground(s) for the deduction and the amount attributable to each ground if there is more than one ground identified. This is commonly referred to as a withholding notice and is a requirement of the HGCRA. Failure to provide a proper withholding notice will normally prevent an Employer from withholding monies due under an Interim Certificate.

4.13.5 If the Employer fails to provide the Contractor with a written notice under either clause 4.13.3 or 4.13.4, then he must pay the amount stated in the Interim Certificate by the final date for payment (i.e. 14 days from the date of issue of the Interim Certificate).

4.13.6 If the Employer fails to properly pay the Contractor on time, then the Contractor is entitled to recover simple interest (at 5% above the Bank of England base rate) on the outstanding monies. Although the Contractor is able to claim and receive interest on this debt the Employer remains in breach of contract for failing to pay the monies at the due time. As the Employer is still in breach of the conditions, the Contractor retains the right to suspend his work or possibly terminate his employment if the breach continues.

Contractor's Right of Suspension

4.14 If the Employer fails to pay the Contractor, by the final date, the full amount of money due (this includes VAT) and there has been no valid withholding notice issued, the Contractor may suspend carrying out the Works. To suspend his performance the Contractor must write to the Employer, with a copy to the A/CA, giving notice of his intentions and the grounds for suspending performance. If the Employer fails to make good the deficiency within 7 days of the Contractor's notice (see clause 1.7) then the Contractor may suspend his performance until he receives full payment from the Employer. This valid suspension by the Contractor is not to be confused with the suspension identified in clause 8.4.1.1, which would allow the Employer to determine the Contractor's employment, nor could the Contractor be charged with failing to proceed regularly and diligently with the Works under clause 8.4.1.2.

Final Certificate – Issue and Payment

4.15.1 The A/CA is required to issue the Final Certificate after the following three events have occurred and must issue it within 2 months of the occurrence of the last event:

4.15.1.1 the end of the Rectification Period for the Works, or (where a project has been broken down into Sections) the end of the Rectification Period for the last Section to be handed over;

4.15.1.2 the date of issue of the 'Certificate of Making Good' issued under clause 2.39, or the issue of the last Certificate of Making Good issued for a Section;

4.15.1.3 the date when the A/CA sent the Contractor copies of the final account and the ascertainment of any loss and/or expense claim. Again, where the project has been broken down into Sections, the relevant date would arise upon the delivery of the last statements of final account and ascertainment of any loss and/or expense (all as clause 4.5.2).

4.15.2 The Final Certificate should provide the following information:

4.15.2.1 the Contract Sum adjusted in accordance with the rules laid down in clause 4.3;

4.15.2.2 the total of the payments included in Interim Certificates plus any advance payment made by the Employer to the Contractor where this option has been used;

The difference between the adjusted Contract Sum and the total payments made under Interim Certificates (plus any advanced payment) will normally result in a payment to the Contractor, although on rare occasions a payment may be due to the Employer because work has been overvalued in the Interim Certificates. The issue of the Final Certificate will not affect any rights that the Contractor may have to recover sums of monies that the Employer has failed to properly pay under earlier Interim Certificates (see clauses 1.10.2 and 1.10.3). For example, the Employer may have withheld monies due in an Interim Certificate without having issued a withholding notice under clause 4.13.3 or 4.13.4, in which case the Contractor still has the right to pursue the Employer for the amount withheld.

Example of a final account layout:

Final Account	£	£
Original Contract Sum		3,420,670.69
Net adjustments to Contract Sum		529,853.31
Loss and/or expense payments		16,548.00
Adjusted Contract Sum		3,967,072.00
Total of sums due under:		
Interim Certificates 1–12	3,500,000.00	
Advance payment	250,000.00	
Sub-total	3,750,000.00	3,750,000.00
Amount due to Contractor		**217,072.00**

4.15.3 To comply with the Housing Grants, Construction and Regeneration Act 1996, the Employer is to give the Contractor a written notice confirming whether he intends to pay the amount stated as being due in the Final Certificate or a different amount. The Employer is also to provide details identifying to what the payment relates and the basis of the calculation. This really just duplicates the Final Certificate, but is a statutory requirement under the Act. The notice is to be given no later than 5 days from the date of issue of the Final Certificate.

4.15.4 The final date by which the monies should be paid is 28 days from the date of issue of the Final Certificate. If the Employer intends to withhold some of the money, he must give the Contractor a notice at least 5 days before the final date for payment. In the notice he must identify the reason(s) for withholding or deducting the money and identify the amount attributable to each reason. If the Final Certificate indicates that there is a balance of payment due from the Contractor to the Employer, then the Contractor is obliged to pay the amount within 28 days from the date of issue of the certificate. There is no provision for the Contractor to challenge this payment through a withholding notice; if the Contractor disagrees with the amount he is required to pay the Employer, he must deal with it under the dispute procedures (see clause 1.10.3).

4.15.5 If the Employer fails to provide the Contractor with a written notice either under clause 4.15.3 or 4.15.4 then he must pay the amount stated in the Final Certificate by the final date for payment (i.e. 28 days from the date of issue of the Final Certificate).

4.15.6 If either the Contractor or Employer fails to pay the amount properly due under the Final Certificate, then simple interest will be added to the debt at 5% above the Bank of England base rate. The fact that interest is claimed on the overdue sum does not mean that the breach (i.e. non-payment of monies due) has been accepted, and the injured party still has the right to take action to recover the late payment.

4.15.7 The balance due under the Final Certificate and any simple interest payable, if any, is treated as a debt due. This just confirms the status of the outstanding monies, and reinforces the fact that action may be taken to recover these monies.

Gross Valuation
Ascertainment

4.16 This clause sets out the procedure to be followed in determining the gross value of the Contractor's work in readiness for the issue of the Interim Certificate. The Gross Valuation is the sum of the values identified in clauses 4.16.1 and 4.16.2, less the values identified in clause 4.16.3. It is important to note that this valuation exercise should not be carried out any earlier than 7 days before the date of issue of the Interim Certificate. For example, if a certificate is due to be issued on 15th March, the earliest date the valuation may be prepared is 8th March.

4.16.1   This clause identifies the items which may be included in the Gross
          Valuation but which are subject to Retention, i.e. their value will be
          reduced by the Retention Percentage stated in the Contract Particulars.
          The relevant items are as follows:

  4.16.1.1   the total value of work properly executed including the value
              of variations and payments due under Fluctuations Option
              C. However, restoration work carried out as a variation under
              Schedule 3, paragraphs B.3.5 or C.4.5.2 is specifically excluded
              from this clause. This is because the Contractor is restoring
              work after damage (e.g. a fire, etc.) and no Retention is to
              be stopped from this money. If the Contract Documents
              have incorporated a priced activity schedule, then that is to
              be used to value the Works. For example, fire insulation to
              steelwork is identified as an activity within the schedule and
              valued at £250,000 – it has been assessed that the Contractor
              has completed 45% of this activity to date, therefore the
              amount to be included in the valuation is £112,500.

              It should be noted that reference is made above to 'the total
              value of work'. For example, the Gross Valuation for interim
              valuation no. 3 would not just be the value of work carried
              out in the previous month but would include the total value
              of work carried out in months one, two and three.
              In the above clause reference is also made to 'work properly
              executed' – this means that no defective work should be
              included in the valuation. However, a Contractor may not
              use this clause to claim that, if works have been included
              in an Interim Certificate, it is evidence that they have been
              accepted and are free of defects (see clauses 1.11 and 3.6).

  4.16.1.2   the total value of materials that have been delivered to site
              but have not yet been built into the Works. It should be
              noted that these materials are only to be included as long
              as they are properly stored and protected and have not
              been brought onto site too early, e.g. engineering bricks
              and reinforcement bars would normally be acceptable if
              delivered to site in month 1 of a 24-month contract, whereas
              a delivery of second fix electrical equipment would not
              normally be acceptable;

  4.16.1.3   materials stored off-site and identified in the Employer's
              list (see clause 4.17). It is important to ensure that all the
              procedures detailed in clause 4.17 are complied with before
              including the value of Listed Items in the valuation.

4.16.2   This clause identifies the items which may be included in the Gross
          Valuation but which are not subject to Retention, i.e. their value must

be paid in full without any deduction for Retention. The relevant items are as follows:

4.16.2.1 any amounts that are to be added to the Contract Sum in accordance with clause 4.4, for example:

payments made to the Contractor for the increase in Option A insurance premiums as a consequence of the Employer's 'early use' of the site (clause 2.6.2),

payment for statutory fees as set out in clause 2.21,

payment of royalties, damages, etc. relating to the use of patent rights under clause 2.23,

payment for tests carried out under clause 3.17,

payments for any third party insurance premium taken out by Contractor in accordance with clause 6.5.3,

payments relating to changes in premium for Terrorism Cover, Insurance Option A,

premiums paid by Contractor where the Employer has failed to take out All Risks Insurance under Insurance Option B,

premiums paid by Contractor where the Employer has failed to take out the appropriate insurance for either existing building or the Works under insurance Option C.

4.16.2.2 loss and/or expense claims payable under clause 3.24.1 or 4.23, or the payment for restoration work etc. that is payable as though it were a variation and is authorised under Insurance Option B or C (see Schedule 3 paragraphs B.3.5 and C.4.5.2),

4.16.2.3 payments made to the Contractor for fluctuations arising under either Fluctuations Option A or B where applicable (i.e. taxes and levies or traditional fluctuation clauses).

4.16.3 This clause identifies situations where monies may be deducted from the Gross Valuation:

4.16.3.1 where deductions arise as a result of any of the following events:

clause 2.10 where setting out errors by the Contractor are not to be rectified,

clause 2.38 where defective work identified during the Rectification Period is allowed to remain,

clause 3.11 where costs have been incurred by the Employer in engaging external contractors to carry out work which the Contractor failed to execute,

clause 3.18.2 where defective work identified during the progress of the Works is allowed to remain, and

      4.16.3.2   where the Employer is entitled to receive credits as a result of fluctuations arising under either Fluctuations Option A or B where applicable (i.e. taxes and levies or traditional fluctuation clauses).

Off-site Materials and Goods

4.17   This clause allows the value of off-site materials to be included in the valuation as long as they have been 'listed' by the Employer (see Definitions clause 1.1) and the following contract terms have been complied with:

    4.17.1   The Listed Items are to comply with the contract, i.e. appropriate quality and specification, etc.

    4.17.2   The Contractor has to have proof that he owns the materials and that they are covered by an appropriate insurance, i.e.:

        4.17.2.1   the Contractor has a legal title to the materials which will allow for the operation of clause 2.25. In other words when the Employer pays for the off-site materials through an Interim Certificate, the legal ownership of the materials will pass to the Employer.

               Note: for further information relating to the ownership of materials and 'Retention of Title' clauses see Aluminium Industrie Vaassen v Romalpa Aluminium Ltd (1976).

        4.17.2.2   the Contractor has to provide proof that the Listed Items are covered by a Specified Perils insurance (see clause 6.8 for a definition of Specified Perils). The policy is to be for the benefit of both the Contractor and Employer and is to be for the full value of the Listed Items. The policy is to be operative from the date when the property in the Listed Items is transferred to the Contractor and remains in force until the Listed Items are delivered to or adjacent to the Works.

    4.17.3   Wherever the materials are stored off-site they must be clearly identifiable by providing the following information:

        4.17.3.1   the name of the Employer, indicating that this is the person for whom they are destined,

        4.17.3.2   the identity and location of the Works where they are to be delivered.

               As well as the two conditions above, the materials are to be clearly set apart from any other materials, or as an alternative, they are to be clearly marked with an appropriate code, etc.

    4.17.4   With reference to 'uniquely identified items' (see JCT 80 Amendment 18 guidelines for a definition and explanation) the Contractor has to have proof that he has an appropriate 'off-site materials' bond where required. The Contract Particulars will specify whether or not a bond is required. If a bond is required it must be provided by a surety that meets the Employer's approval, be for the amount stated in the Contract Particulars and incorporate the terms set out in Schedule 6, Part 2.

4.17.5   For Listed Items that are not uniquely identified the same terms as above apply, except that in this instance a bond is compulsory.

Retention
Rules on Treatment of Retention

4.18   This clause sets out the rules relating to Retention:

4.18.1   'The Employer's interest in the Retention is fiduciary', means that the Employer holds the Contractor's Retention in trust and acts as a trustee for this money. The Employer is not obliged to invest the Retention money.

4.18.2   At the date of each Interim Certificate the A/CA (or QS) is to prepare a Retention statement identifying the amount of Retention being withheld from the Contractor. The A/CA must send the Retention statement to the Employer, with a copy to the Contractor. This procedure ensures that both parties are fully aware of the amount of Retention that is being held back by the Employer.

4.18.3   Where the Employer does withhold Retention, the Contractor may ask for the Retention to be placed in a separate and properly designated bank account (so that the Retention monies may be readily traced and identified). As a result of such a request, the Employer will have to place the Retention into this account after each Interim Certificate and confirm his actions by providing the A/CA with an appropriate certificate (with a copy to the Contractor). Any interest earned from this account will go to the Employer. This clause does not apply where the Employer is a Local Authority.

The purpose of this clause is to give the Contractor the opportunity to safeguard the Retention monies in case a 'private' Employer gets into financial difficulties. However, the separate bank account will only be set up if the Contractor requests it.

(See Rayack Construction v Lampeter Meat Co. 1979;

Wates Construction (London) Ltd v Franthom Property Ltd, 1991;

J F Finnegan Ltd v Ford Sellar Morris Developments Ltd, 1991;

MacJordan Construction Ltd v Brookmount Erostin Ltd, 1991;

GPT Realisations v Panatown, 1992)

4.18.4   If the Employer deducts monies from the Contractor's Retention, he must inform the Contractor of the amount to be deducted by reference to the latest Retention statement.

Retention Bond

4.19 This is an optional clause and, if it applies (see Contract Particulars), there will be no deduction of Retention from the Interim Certificates. As a result certain procedures will follow:

4.19.1 The provisions under clauses 4.10.1 and 4.20 that allow the deduction of Retention from monies due to the Contractor are suspended. Having said that, it is still necessary for the A/CA or QS to calculate the amount of Retention and prepare a statement of Retention in accordance with the Retention rules. The reason for this apparent anomaly is to inform the Employer (also the Contractor and surety) as to what Retention would have been available under the normal rules.

4.19.2 The Contractor is required to obtain and maintain a Retention bond in accordance with Schedule 6 Part 3. The bond is to be to the benefit of the Employer and must comply with the details incorporated in the Contract Particulars regarding the maximum aggregate sum and the expiry date of the bond. The bond must be provided by a surety acceptable to the Employer and must be supplied to the Employer on or before the Date of Possession.

4.19.3 If the Contractor fails to provide or maintain the bond, then the normal Retention rules will apply and Retention will be deducted from the Interim Certificates issued after the breach. If and when the Contractor provides the necessary bond, the Employer will release any Retention through the next Interim Certificate.

4.19.4 If, during the progress of the project, the amount of Retention that could have been deducted under the normal rules exceeds the aggregate sum stated in the bond, then the Contractor shall increase the value of the bond, or the Employer may deduct the value of the Retention that exceeds the bond. This situation may arise because the Contract Sum is increased as a result of extensive variations or the operation of formula fluctuations.

4.19.5 In the situation where the Employer has required the Contractor to provide a performance bond, and a Contractor default arises which could be claimed under either the performance bond or the Retention bond, then the Employer must first seek recourse under the Retention bond.

Retention – Rules for Ascertainment

4.20 This clause identifies the means by which the Retention is to be calculated, i.e. a percentage deduction from the amount included in the Interim Certificate for the items identified in clause 4.16.1.

4.20.1 The Retention will be 3% unless a different rate has been inserted in the Contract Particulars.

4.20.2 Full Retention may be deducted from:

4.20.2.1 the Works that have not reached practical completion (or practical completion of a Section). Although not mentioned in this clause, it is important to be aware that a statement of partial possession will also affect the amount of Retention that may be deducted;

4.20.2.2 amounts included for materials on site and Listed Items that are off-site.

4.20.3 Half Retention is to be deducted from the above after the issue of a Practical Completion Certificate for the Works (or a Section) and where a Making Good Certificate has not been issued for the Works (or Section or, with reference to partial possession, a certificate for making good under a Relevant Part).

A simple explanation of the above rules is that full Retention is withheld from the appropriate items until the issue of a Practical Completion Certificate or a statement of partial possession, when half the Retention is released. The remaining Retention is released upon issue of the Certificate of Making Good, or Certificate of Making Good under a Relevant Part.

Fluctuations
Choice of Fluctuation Provisions
This section determines which party will be largely responsible for carrying the risks of future price changes in the construction work.

4.21 The various procedures for dealing with fluctuations are set out in Schedule 7. There are three possible choices of fluctuation clauses and the Employer, when completing the Contract Particulars, should specify which clause is to apply to this contract. If no Fluctuation Option is identified in the Contract Particulars, then by default Option A, identified below, is to apply. The choice of fluctuation procedures available to the Employer are listed below:

Fluctuations Option A: i.e. fluctuations in the cost of the construction work incurred as a result of changes in levies and taxation;

Fluctuations Option B: as Option A plus an allowance for fluctuations in the cost of materials and labour as a result of changes in market prices;

Fluctuations Option C: a similar level of fluctuations recovery is allowed as in Option B, but the calculations are based upon a set of 'formula rules' and published indices.

Non-Applicable to Schedule 2 Quotation
4.22 Work or variation work carried out by the Contractor under a Schedule 2 Quotation is excluded from any fluctuations adjustment.

Loss and Expense
Matters Materially Affecting Regular Progress
4.23 To initiate the procedures contained within this clause, the Contractor must first write (i.e. make an application) to the A/CA stating that he has incurred (or is likely to incur in the future) direct loss and/or expense which will not be

recovered under any other clause within the contract. In this application the Contractor may provide details regarding the value of the claim, although this is not obligatory (but see clause 4.23.3). The loss and/or expense must result from the Employer having deferred possession of the site under clause 2.5 (where this clause is operative) or from the Works being disrupted and 'materially affected' by a Relevant Matter identified in clause 4.24. Upon receipt of the Contractor's application and if the A/CA considers the claim to be valid (i.e. the Contractor has or will incur a direct loss and/or expense that is reimbursable under this clause) then he must assess the value of the claim, or he may delegate this task to the QS.

When assessing the validity of the Contractor's application the A/CA must consider, amongst other things, whether the Contractor is prevented from making a claim through the operation of other clauses within the contract, e.g. clause 2.20 identifies certain events for which the Contractor is unable to claim loss and/or expense under clause 4.23.

To ensure the proper operation of clause 4.23 the Contractor should be aware of the following:

4.23.1   the Contractor is required to submit the above application to the A/CA as soon as he is aware (or should have been reasonably aware) that the regular progress of the work has been, or is likely to be, affected;

A Contractor needs to be vigilant to ensure he complies with the requirement to provide a prompt notice. If the Contractor fails to submit a timely notice this could potentially result in the A/CA refusing to consider the application and consequently not operating the clause 4.23 procedures. In such a situation the Contractor would have to consider taking legal action to recover his loss and expense with all the attendant costs, delays and uncertainties;

4.23.2   the A/CA may request the Contractor to provide further information so as to reasonably allow the A/CA to form an opinion of whether the Contractor has, or is likely to incur, direct loss and/or expense;

4.23.3   the A/CA or QS may request the Contractor to provide reasonably sufficient details of the loss and/or expense to allow the claim to be evaluated.

Relevant Matters

4.24   This is an important clause as it identifies the Relevant Matters that may entitle the Contractor to raise a claim under clause 4.23. Below is a list of the Relevant Matters:

4.24.1   variations, which also includes matters or instructions which are to be treated as though they were variations (e.g. clause 2.14.3 the correction of errors in the Contract Documents and Schedule 3, paragraph B.3.5 restoration work, etc. are instances where work is to be treated as a variation).

It is important to note that a Schedule 2 Quotation and any relevant variation is excluded from the list of Relevant Matters because the

Contractor is required to include an allowance for loss and/or expense in his Schedule 2 Quotation. Also, any variations to Schedule 2 Quotation works must include the value of any loss and/or expense (see clause 5.3.3).

4.24.2 the following instructions issued by the A/CA:

4.24.2.1 clause 3.15 – the postponement of any of the Works; and clause 3.16 – an instruction for the expenditure of a Provisional Sum. However, an instruction for the expenditure of a 'defined' Provisional Sum is specifically excluded as a Relevant Matter. The reason for this exclusion is that a defined Provisional Sum provides the Contractor with sufficient information about the work to be executed and he is deemed to have made full allowance in his programme, etc. for carrying out this work;

4.24.2.2 clause 3.17 – opening up and testing the Works and making good, unless the tests proved the work to be defective;

4.24.2.3 clause 2.15 – a discrepancy or divergence between the Contract Drawings, Contract Bills and/or other documents.

4.24.3 suspension of the Works by the Contractor under clause 4.14, as long as the suspension was not unreasonable;

4.24.4 where the Contractor has carried out work covered by an Approximate Quantity in the bill of quantities and this Approximate Quantity proves not to be a reasonably accurate forecast of the quantity of work actually executed. For example, a bill of quantities may have contained an Approximate Quantity for 10m of land drains, whereas the Contractor was eventually required to install 1,000m of land drains. This additional work could easily lead to a disruption to the progress of the Works, and as the original estimate of 10 metres was clearly inaccurate, the Contractor would have strong grounds for a claim;

4.24.5 this sub-clause reviews the situation where an Employer is responsible for causing a Contractor to incur loss and/or expense. However, instead of trying to identify all the individual Employer defaults that may give rise to a claim for loss and/or expense, this clause is written in fairly broad terms.

A Relevant Matter occurs where the regular progress of the Works is disrupted by '*any impediment, prevention or default*' of the Employer, A/CA, QS or any Employer's Persons (see Definitions clause 1.1). The Contractor's ability to claim under this Relevant Matter may be reduced or negated if it can be shown that the Contractor (or Contractor's Persons, e.g. sub-contractors) through his acts or omissions was partly or totally responsible for causing the Employer default.

Amounts Ascertained – Addition to Contract Sum

4.25 As and when the A/CA or QS calculates the amounts due to a Contractor under a claim for loss and/or expense, that amount is to be added to the Contract Sum, i.e. in effect the amount is to be included in the next Interim Certificate.

Reservation of Contractor's Rights and Remedies

4.26   The presence of clauses 4.23 to 4.25 does not prevent the Contractor from recovering loss and/or expense by other means. Clauses 4.23 to 4.25 provide a procedure within the contract conditions which allows a loss and/or expense event to be resolved but, if the Contractor chooses, he could pursue an action through the courts for breach of contract – although this approach could be expensive and time consuming.

Section 5: Variations

General
Definition of Variations

5.1　This clause provides a broad definition of what is meant by the term 'variation', i.e.:

　　5.1.1　changes or modifications to the design as well as the quality or the quantity of the Works, which may include any of the following:

　　　　5.1.1.1　the addition or omission of work or substitution of work – most variations will tend to come under this heading. This clause allows the A/CA to instruct the Contractor to carry out more work than was originally allowed for in the Contract Documents, or he may instruct that certain works are not to be executed, i.e. they are omitted from the Works. The A/CA may also change the original design, for example change a flat roof to a pitched roof;

　　　　5.1.1.2　the change in specification of any goods or materials, e.g. the alteration of half-hour fire check doors to one-hour fire check doors;

　　　　5.1.1.3　the removal of work or materials which are in accordance with the Contract Documents, i.e. work that is not defective. This may arise when the Contractor has carried out some work in accordance with the Contract Documents but, because of a design change, this work is no longer required. The Contractor will be paid to dismantle and remove the work.

　　5.1.2　changes to the Employer's obligations, i.e. the Employer (through an instruction of the A/CA) is able to impose obligations or restrictions on the Contractor with specific reference to:

　　　　5.1.2.1　access to the site, or the specific use of parts of the site;

　　　　5.1.2.2　limitation of working space. This refers to the general use of the site and does not refer to 'working space' as defined by the SMM7;

　　　　5.1.2.3　limitation of the Contractor's working hours on site;

　　　　5.1.2.4　carrying out the work in a specific order.

　　　　　　Having imposed such obligations or restrictions, the Employer (through the A/CA) may subsequently add to, omit or alter any of the obligations. Similarly, where the Employer has imposed obligations or restrictions on the Contractor through the Contract Bills or Employer's Requirements, it is possible to add to, omit or alter the obligations or restrictions.

　　　　　　Because of the serious implications that such a variation may have on a Contractor's site operations, the Contractor has a right to submit a reasonable objection to a 5.1.2 variation (see clause 3.10.1).

Valuation of Variations and Provisional Sum Work

　　5.2.1　This clause identifies how variations work should be valued. Initially it identifies the type of work that is to be valued under this clause, e.g.:

5.2.1.1   variations issued by the A/CA as an instruction (see clause 3.14 – this identifies that it is the role of the A/CA to order variations and they are to be issued as an A/CA's instruction);

5.2.1.2   all work that is to be treated as a variation in accordance with these conditions, e.g. see clause 2.14.3 and Schedule 3, paragraph C.4.5.2;

5.2.1.3   all work carried out by the Contractor in response to an instruction from the A/CA as to the expenditure of Provisional Sums included in the Contract Bills or Employer's requirements;

5.2.1.4   all work carried out by the Contractor that is covered by an Approximate Quantity within the Contract Bills or Employer's requirements.

The value of the variation is to be an amount that is agreed between the Employer and Contractor. Where the two parties cannot reach an agreement on the value of the variation, the task will be passed to the QS, unless the Employer and Contractor decide on an alternative approach, e.g. they may appoint an external consultant or specialist. If the QS is to value the variation, he must do so in accordance with the valuation rules contained in clauses 5.6 to 5.10. Where the variation relates to CDP work the QS must value this in accordance with clause 5.8.

5.2.2   The above rules do not apply to a variation where the Contractor has supplied a Schedule 2 Quotation which has been accepted, nor do they apply to a variation to work carried out under an accepted Schedule 2 Quotation. The value of the Schedule 2 Quotation is dealt with in Schedule 2 paragraph 3.1.2.2, and any subsequent variation is dealt with under clause 5.3.3.

Schedule 2 Quotation

5.3.1   This clause will only apply where it is stated in the A/CA's instruction that the Contractor is to provide a quotation for the variation in accordance with the procedures set out in Schedule 2. There is also a proviso that the Contractor must be given sufficient information to allow him to provide a quotation (see Schedule 2 paragraph 1.1 and footnote [62]). The Contractor can notify the A/CA that he is not prepared to submit a Schedule 2 Quotation but he must normally do so within 7 days of receiving the instruction. In his instruction the A/CA may stipulate that the Contractor has a longer period than 7 days period to consider whether or not he wishes to provide a Quotation. Alternatively the 7-day period may be extended with the agreement of the A/CA and Contractor.

5.3.2   Where the Contractor properly notifies the A/CA that he is not prepared to submit a Quotation, he is not obliged to comply with the variation instruction. The variation must not be carried out unless the A/CA issues another instruction stating that the variation work to be carried out and that it will be valued in the normal manner, i.e. by a valuation.

5.3.3   If the A/CA issues an instruction to vary the work of an accepted Schedule 2 Quotation, the variation is not to be valued under clause 5.2. The QS is to value the variation on a 'fair and reasonable basis' – taking into account the content of the relevant Quotation. The QS must also include the value of any direct loss and/or expense incurred by the Contractor as a result of complying with this instruction.

Contractor's Right to Be Present at Measurement

5.4   Where it is necessary for the QS to measure work prior to valuing a variation, he is to inform the Contractor and allow him to be present so he may take his own measurements and prepare any notes.

This situation could arise where the QS has becomes responsible for valuing the works under 5.2.1 and the procedure should ensure that the Contractor and QS are generally in agreement over the quantity and scope of the variation works.

Giving Effect to Valuations, Agreements etc.

5.5   The value of variations is to be added to or deducted from the Contract Sum (see clause 4.4 to appreciate the implication of this statement).

The Valuation Rules

If the QS is responsible for valuing variations, then he must carry out the task in strict compliance with the following rules:

Measurable Work

5.6.1   The following valuation rules apply to variations for additional or substituted work *which can be properly valued by measurement*. The rules also apply to work executed by the Contractor and which is allowed for in the Contract Bills by way of an Approximate Quantity (all Approximate Quantities have to be remeasured). These rules do not apply to variations relating to CDP work; they are dealt with under clause 5.8. The rules are as follows:

5.6.1.1   additional or substituted work that is basically the same as work already measured in the Contract Bills (i.e. it is of the same character and carried out under the same conditions) will be priced at bill rates. An additional requirement is that the variation does not *significantly* change the quantity of the work set out in the bill of quantities;

5.6.1.2   where additional or substituted work is of a similar character to that contained within the Contract Bills, but the conditions under which the work is executed are different and/or the variation significantly changes the quantity of similar work set out in the Contract Bills, then the bill rate will provide a basis for the valuation. The original bill rate should be analysed and adjusted to reflect the cost implications incurred by the Contractor because of the changed conditions and/or significant change in quantity;

(See Wates Construction Ltd v Bredero Fleet Ltd, 1993; Henry Boot Construction Ltd v Alstom Combined Cycles, 2000)

5.6.1.3 where the additional or substituted work is not of the same character as that set out in the Contract Bills, then a new rate has to be calculated on the basis of using fair rates and prices. It is generally accepted that the term 'fair rates and prices' implies that any new rate should reflect the level of pricing used by the Contractor when originally pricing the Contract Bills, i.e. if the Contractor only allowed for 5% profit and overheads in the Contract Bills then the same will apply to the new rate. This is simple in theory but not so easy in practice;

(See Norwest Holst Construction Ltd v Co-operative Wholesale Society Ltd, 1997)

5.6.1.4 on remeasuring an Approximate Quantity (remember that all Approximate Quantities must be remeasured), the work is to be valued at the original bill rate as long as the Approximate Quantity was a reasonably accurate forecast of the actual work executed;

5.6.1.5 where an Approximate Quantity was not a reasonable forecast of the work executed, the bill rate will form the basis of the valuation. The bill rate will be adjusted to reflect the cost implications of the difference in quantity between the Approximate Quantity and the quantity actually installed. For example, if a far larger amount of work was executed than allowed for in the Approximate Quantity, it could be argued that the Contractor may have benefited from improved trade discounts and increased productivity. If this argument were accepted, then the original bill rate would be reduced to reflect these issues.

The rules in 5.6.1.4 and 5.6.1.5 will only apply to Approximate Quantities where there has been no change in the quality, specification, etc. of the item, therefore these two rules address purely how the overall quantity may affect the original bill rate.

5.6.2 Where a variation requires an item to be omitted from the Works, then that item must be omitted at the bill rate. However, where an omission relates to CDP work, it must be dealt with in accordance with clause 5.8.

5.6.3 The following rules are to apply to any valuation carried out under 5.6.1 and 5.6.2:

5.6.3.1 any variation work that has to be measured, must be measured in accordance with the measurement rules originally used in the preparation of the Contract Bills. This will normally mean the use of the Standard Method of Measurement 7th edition (see Definitions clause 1.1) but the Contract Bills should be checked to see if there has been any departure from this normal convention;

5.6.3.2 if the Contract Bills are subject to a percentage adjustment, e.g. if a mathematical error was found in the Contract Bills during the tender evaluation, this may result in an overall percentage adjustment being applied to the bills to correct the error and to ensure the Contract Sum remains the same. This may lead to a small percentage deduction (or addition) on the summary page, e.g. plus 0.015% – and the same percentage adjustment would be added to all variations valued at bill rates;

5.6.3.3 if a variation results in an increased or decreased expenditure on a preliminaries item, then this must be taken into account when pricing the variation, e.g. additional work may require the use of plant, scaffolding or hoisting equipment beyond what had been allowed for in the preliminaries. This rule does not apply to work carried out against a defined Provisional Sum as the Contractor is deemed to have made full allowance in his preliminaries for this provisional work at the tender stage.

Daywork

5.7. If it is thought that a variation for additional or substituted work cannot be properly valued by measurement, then the following rules may be used:

5.7.1 the work may be valued as a 'daywork' in accordance with the 'Definition' prepared by the RICS and the Construction Confederation. The Definition to be used is the one that was current at the contract Base Date. Any percentage additions the Contractor wished to apply to the daywork should have been written into the Contract Bills by the Contractor at the tender stage;

5.7.2 where the variation work is of a specialist nature and the RICS, in conjunction with the appropriate organisation, has an agreed and published daywork definition for that work, then the specialist work may be valued in accordance with that definition. The RICS currently have agreed definitions with the Electrical Contractors' Association and the Heating and Ventilating Contractors' Association.

If it is envisaged that an item of work is to be valued on a daywork basis, then the Contractor must provide worksheets giving details of the workmen's names and time spent on the work, as well as plant and materials used. The worksheet is to be submitted to the A/CA (or his authorised representative) for verification, i.e. for the A/CA to acknowledge that in his opinion the worksheet is reasonably accurate. The Contractor must submit the worksheet no later than the end of the week following the execution of the work.

Contractor's Designed Portion – Valuation

5.8 Where a Contractor undertakes CDP work he is involved in additional costs and documentation not normally associated with a traditional procurement contract, e.g. design work, professional indemnity insurance, CDP Analysis, etc. As a consequence variations relating to CDP work are treated on a slightly different basis from the normal variations. Such variations are dealt with in accordance with the following rules:

5.8.1 when valuing a variation it is necessary to take into account the cost of any associated design work;

5.8.2 where a variation relates to additional or substituted work, the valuation should reflect the values of similar work contained in the CDP Analysis. If the variation work is carried out under different conditions, or there is a significant change in the quantity of the work, these aspects must be taken into account in the valuation. Where a variation requires work to be executed which is of a different character, then it should be valued on the basis of a fair valuation.

A problem with trying to apply the above rule is that the CDP Analysis may provide only a very limited breakdown of the Contractor's costs, making it very difficult to identify any prices that are relevant to the variation work.

5.8.3 where the variation requires an omission of CDP work, the omission will be valued in accordance with the values in the CDP Analysis.

Once again, this may be a problem when a small part of an element of CDP work is omitted and the CDP Analysis provides only a very general breakdown of that element.

5.8.4 when valuing a CDP variation, the QS should apply the rules from clauses 5.6.3.2, 5.6.3.3, 5.7 and 5.9, where appropriate.

Change of Conditions for Other Work

5.9 If, as a consequence of one of the following events, there is a substantial change in the conditions under which *other* work is executed, it is possible to consider whether that other work should be re-priced. The events are:

5.9.1 compliance with any instruction requiring a variation, or

5.9.2 compliance with any instruction to execute work against an undefined Provisional Sum, or

5.9.3 compliance with any instruction to execute work against a defined Provisional Sum where the work executed is different from the description provided in the Contract Bills, or

5.9.4 the execution of work covered by an Approximate Quantity, where the quantity of work differs from the Approximate Quantity allowed for in the Contract Bills.

If, as a result of any of the above, there is a substantial change in the conditions in which other work is executed, then the *other* work is to be treated as though it were the subject of a variation order and may be valued in accordance with clause 5, e.g. the work may be revalued in accordance with 5.6.1.2. For example, a situation may arise where the A/CA issues a variation altering the specification of the steel frame and, as a consequence, the fixing of the cladding panels (i.e. other work) becomes a much more complex task. Under such circumstances the cladding panels may now be treated as though they were covered by a variation and be valued in accordance with 5.6.1.2. This clause may also be used in relation to CDP work.

Additional Provisions

5.10.1 Where a valuation does not relate to the execution of additional or substituted work, or the omission of work, or where the valuation of any work or liabilities directly associated with a variation cannot be reasonably carried out under the rules of 5.6 to 5.9, the variation shall be valued on the basis of a fair valuation. For example, a variation issued under clause 5.1.2 may fall within this rule. A change in obligations or restrictions would be difficult to value using the normal valuation rules and a 'fair valuation' is perhaps the only reasonable approach to take.

5.10.2 A variation can sometimes result in the disruption of the regular progress of the work, causing the Contractor to incur loss and/or expense as a result. No allowance should be made for this loss and/or expense when valuing the variation, provided the loss and/or expense may be recovered under another clause within the contract. For example, loss and/or expense incurred as a result of a normal variation would be recovered through the procedures laid down in clause 4.24.1. However, a variation to Schedule 2 Quotation work could not be recovered under clause 4.24.1 and should be dealt with through clause 5 (see clause 5.3.3).

Section 6: Injury, Damage and Insurance

Injury to Persons and Property
Liability of Contractor – Personal Injury or Death

6.1 Where a person is injured or killed as a direct consequence of the Works being carried out then the Contractor will indemnify the Employer against any costs, liability, claim, expenses, etc. that the Employer may incur relating to such personal injury or death. However, the Contractor will not be responsible for indemnifying the Employer if the injury or death was caused by the negligence of the Employer or persons employed by him (see Definition 'Employer's Persons' clause 1.1).

Liability of Contractor – Injury or Damage to Property

6.2 The Contractor is liable to indemnify the Employer where real or personal property is damaged (e.g. other buildings or a motor vehicle) as a direct consequence of the Works being carried out. This indemnity does not include the Works while they are in the Contractor's possession, nor existing property insured by the Employer under Insurance Option C (see Schedule 3 paragraph C.1). The Contractor is liable to indemnify the Employer only where the damage has been caused by the Contractor's negligence, etc. or by persons employed or engaged by the Contractor (see Definition 'Contractor's Persons' clause 1.1); damage caused by the negligence of the Employer, or Employer's Persons is excluded from the Contractor's liability.

Note: Employer's Persons are defined as persons engaged, employed or authorised by the Employer but there are a number of parties who are specifically excluded from this definition, i.e.: the Contractor, the A/CA, the QS and any Statutory Undertaker. An example of a person who would fall within the definition of Employer's Person may be found in clause 3.23, i.e. a third party who is authorised to excavate and examine any antiquities discovered on site.

Injury or Damage to Property – Works and Site Materials Excluded

6.3.1 This clause is a confirmation that the term 'property real or personal' does not include property such as the Works, work executed or materials on site. However, this only applies until:

6.3.1.1 the date of issue of the Practical Completion Certificate, or
6.3.1.2 the date of termination of the Contractor's employment.

After any of the above events, and no matter which occurs first, the Contractor is potentially liable for damage caused to the Works, materials on site, etc.

An explanation of clause 6.3.1 is that the Works and materials, etc. are already covered by alternative insurance provisions (Insurance Option A, B or C) and are therefore excluded from the Contractor's liability in clause 6.2. However, after practical completion or termination of the Contractor's employment, the obligation to maintain insurance cover under Insurance Option A, B or C ceases.

6.3.2 This is an explanation to the effect that the Contractor will not be liable under clause 6.2 for damage to any work executed in a Section prior to a Section Completion Certificate being issued. Once the certificate has been issued the Contractor is potentially liable for damage caused to the work in that Section.

6.3.3 Following a statement of 'partial possession', the Relevant Part of the Works taken over by the Employer ceases to be part of the Works or work executed, therefore, after the Relevant Date (but not before), the Contractor is potentially liable for damage to this property caused by his negligence.

Insurance against Personal Injury and Property Damage
Contractor's Insurance of His Liability

Clause 6.4 outlines the insurance cover which the Contractor is required to have in place to provide a financial backing for his indemnities identified in clauses 6.1 and 6.2, and the optional insurance that may be required under clause 6.5.

6.4.1 The insurances taken out by the Contractor to cover the indemnities he provides to the Employer in clauses 6.1 and 6.2 do nothing to limit or remove the Contractor's obligation to indemnify the Employer under clauses 6.1 and 6.2. For example, if the insurances have a financial limit or exclude certain liabilities, the Contractor is still liable to indemnify the Employer and cannot claim that his liability to indemnify the Employer is reduced in line with the insurance policy provided. The insurances taken out and maintained by the Contractor must comply with the following:

6.4.1.1 The insurance for injury to persons employed by the Contractor (where injured in the course of their employment) is to comply with all relevant legislation (e.g. Employer's Liability [Compulsory Insurance] Act).

6.4.1.2 Insurance for other claims under 6.4.1 should cover the Contractor's liability to indemnify the Employer under clauses 6.1 and 6.2. The minimum amount of financial cover required (per event) is to be inserted in the Contract Particulars by the Employer. It is important to note that the amount stated in the Contract Particulars is a minimum requirement and it is up to the Contractor to decide whether it is adequate or whether a greater amount of cover should be obtained (see footnote [52]). If a claim is submitted which exceeds the financial limit of the insurance, the Contractor will be liable for any shortfall.

The wording used in this section of the contract reflects the fact that the Employer and Contractor are classified as 'joint tortfeasors', i.e. an Employer may be sued for a death or injury caused by the Contractor. The Employer is covered for this event by the Contractor's indemnity and insurance.

6.4.2 The Employer can make reasonable requests to the Contractor to provide the A/CA with documentary evidence that the above insurances are in place. This allows the Employer, as and when he thinks it appropriate, an opportunity to inspect the documentation to ensure the necessary insurance is in place and is being maintained. The Employer may, at any time, reasonably request the Contractor to send to the A/CA the insurance policies and premium receipts so that the Employer may inspect them. An Employer may wish to re-inspect the documents on or after an annual renewal date. It is presumed that the documents are to be returned after inspection.

6.4.3   If the Contractor fails to take out or maintain the clause 6.4.1 insurances, the Employer *may* take out similar insurances to protect himself. The cost of the insurance(s) may be recovered from monies due to the Contractor, or charged to the Contractor, e.g. the monies may be recovered from an interim payment through the issue of a withholding notice, or the Contractor may be billed for the amount due.

Note should be made that the Employer is not obliged to take out insurance where the Contractor is in default; it is an option open to the Employer. If the Employer does take out his own insurance, it is only to cover his liabilities, it will not provide cover to the Contractor.

Contractor's Insurance of Liability of Employer

6.5   This is an optional clause which details the insurance cover for damage that may be caused to nearby property where the damage has not been caused by the Contractor's negligence (if the Contractor has not been negligent, then this damage will not be covered by the Contractor's indemnity in clause 6.2, see Gold v Patman & Fotheringham Ltd, 1958). If the Employer thinks he may require this insurance, then this must be stated in the Contract Particulars. The insurance is to be taken out by the Contractor only if instructed by the A/CA. The insurance is to be in the names of the Employer and Contractor and provide the amount of cover stated in the Contract Particulars. In this instance, if the amount of cover is inadequate to cover a claim then any shortfall will be the Employer's liability. The insurance should be maintained until the end of the Rectification Period (see footnote [53]). The insurance is to provide cover for damage to property caused through any collapse, subsidence, heave, vibration, weakening or removal of support or lowering of ground water caused by the carrying out of the Works. However, there is a considerable list of exclusions (based on the model exclusions produced by the ABI) from the cover to be provided, i.e.:

6.5.1.1   damage caused by the Contractor's negligence. This would obviously be covered by the Contractor's indemnity provided under clause 6.2;

6.5.1.2   design errors. This would appear to apply to design errors, for which the A/CA is responsible, and to Contractor design errors in any CDP work. Both of these instances should be covered by the defaulting party's professional indemnity insurance;

6.5.1.3   damage which can seen to be reasonably foreseeable. For example, on some projects it may be envisaged that pile driving operations will almost certainly cause disturbance to an adjoining property. Any such foreseeable damage would not be covered by this insurance;

6.5.1.4 existing property insured by the Employer under Insurance Option C, Schedule 3 paragraph C.1;

6.5.1.5 the Works and Site Materials, as these will already be covered by the Works All Risks insurance. However, work taken over by the Employer following a Practical Completion Certificate (or Section Completion Certificate) will not be covered because the All Risks policy will have lapsed upon the date of issue of these certificates. The same situation should arise where the Employer takes over part of the Works through partial possession but the contract remains silent about this event;

6.5.1.6 damage caused through war, hostilities etc.;

6.5.1.7 damage caused directly or indirectly from an excepted risk (see clause 6.8 for a list of excepted risks);

6.5.1.8 damage caused directly or indirectly by contamination or pollution during the course of the insurance, i.e. problems of pre-existing pollution or pollution events that have gone unnoticed are excluded; for example, a long-term leak from an oil storage tank. However, contamination or pollution that occurs as a sudden identifiable accident during the course of the insurance will be covered, such as the accidental rupturing of an oil storage tank;

6.5.1.9 costs, etc. incurred by the Employer from a breach of contract. However, if any of those costs would have been payable by the Employer even though a contract was not in place, these will be covered.

6.5.2 The above insurance is to be taken out with a company approved by the Employer. The policy and premium receipts are to be sent to the A/CA, who will forward them to the Employer, who will retain them.

6.5.3 The cost to the Contractor of taking out and maintaining the insurance will be added to the Contract Sum (see clause 4.4).

Excepted Risks

6.6 The Contractor is not liable to indemnify the Employer or take out insurance to cover damage caused by an 'excepted risk' (see clause 6.8 definitions to appreciate what comprises an excepted risk). Consequently the Contractor's liability to indemnify the Employer under clauses 6.1 and 6.2 is modified by this clause, as is his obligation to provide covering insurance under clause 6.4. Insurance companies will not provide cover for excepted risks and these risks are usually covered by statutory legislation and compensation.

Insurance of the Works

This section details the insurance cover required for the works and site materials. It determines whether the Contractor or Employer will be responsible for the insurance, and deals with the slightly more complicated situation where works are being carried out in, or adjoining, an existing building. The footnotes to this part of the contract provide more information as detailed below:

Footnote [54]   Insurance Options A and B are for use where the work is to be the erection of a new building. Where Option A is chosen, the Contractor is responsible for taking out a Joint Names Policy for all-risks insurance, and where Option B is chosen, the Employer takes out the Joint Names Policy for all-risks insurance. Where the work comprises alterations to an existing structure and/or an extension to an existing structure, Option C must be used. Under Option C the Employer is responsible for taking out a Joint Names Policy for all-risks insurance for the Works, and a Joint Names Specified Perils insurance for the existing building and for the contents owned by the Employer, or for which he is responsible. It is advised that some Employers may have difficulty in obtaining Joint Names cover for the Option C insurances and in particular the Specified Perils insurance cover. In such a situation the contract will have to be amended, e.g. Option C should not be used and an alternative approach will have to be adopted. The JCT Guide[2] provides some initial advice on this matter, see paragraphs 70, 82 and 83.

Footnote [55]   Clause 6.8 defines the level of cover that the parties must provide when arranging the All Risks Insurance under this contract. Attention is drawn to the fact that there is not a standard All Risks policy that is issued by all insurers, the wording of an All Risks policy and its cover may vary from insurer to insurer. It is a requirement that Terrorism Cover is obtained as part of an All Risks policy although because of the nature of this risk there may be times when this is not possible. In such a situation the parties are advised to enter into discussions with their insurance advisers and refer to the JCT Guide. The Guide provides very limited advice on this matter apart from informing the parties to be aware of what level of Terrorism Cover, if any, is being provided by a policy and to be aware of the alternative approaches detailed in clause 6.10 for when Terrorism Cover is withdrawn during the progress of the Works.

Footnote [56]   Specific attention is drawn to the exclusion from the All Risks definition of loss or damage caused through defective design, materials or workmanship. An All Risks policy that excludes the insurer's liability beyond that defined in clause 6.8 (b) is not acceptable under this contract.

Insurance Options

6.7   The Employer must, in the Contract Particulars, state which insurance clause is to apply to this contract, i.e. Insurance Option A, B or C. Further details relating to these Insurance Options may be found in Schedule 3.

Footnote [54] provides further information concerning the Insurance Options.

Related Definitions

This clause provides definitions for some of the terms that are used in the insurance clauses.

6.8 The following definitions are to apply to the text found in Schedule 3, i.e. the Insurance Options. The definitions will also apply to the contract conditions, where relevant. The definitions are as follows:

All Risks Insurance: This is an insurance to cover the Works and Site Materials from loss or damage. The insurance is also to cover consequential costs such as the removal and disposal of damaged work and materials and the provision of temporary supports. However, the term All Risks can be misleading as there are a number of specific events that are excluded from such a policy, as identified in 6.8 (a) to (c):

6.8(a) property which has become defective as a result of wear and tear, obsolescence, deterioration, rust or mildew (these would not be viewed by an insurer as accidents);

6.8(b) work or property that is lost or damaged as a result of an inherent design defect, inappropriate specification, material defects or defective workmanship. This exclusion includes any other work or property that is reliant on the defective work for support or is attached to the defective work. For example, a Contractor constructs an in-situ reinforced concrete beam with inadequate reinforcement (i.e. defective workmanship); the beam collapses, as does the wall it was supporting. None of this damage is covered by the All Risks Insurance. However, the terrazzo floor finish damaged by the falling masonry is covered by the insurance as it was not supported by or attached to the defective beam. Also see advice contained in footnote [56].

6.8(c) (i) loss or damage caused by, or arising from, war, invasion, civil war, etc. Also loss or damage caused by an order of the government, or of the public or local authority, e.g. the demolition of a wall to alleviate flooding and damage to nearby premises.

Note: *de jure* relates to a government that has been lawfully elected, *de facto* refers to a government that is actually in power, although it may not have been lawfully elected.

6.8(d) (ii) loss or damage caused by disappearance or shortage where this has only become evident following an inventory, or where it is not possible to associate the disappearance or shortage to an identifiable event. The apparent disappearance of bricks through excessive wastage would not be covered, whereas the disappearance of packs of bricks following a break-in to the site would be covered.

6.8(e) (iii) loss or damage caused by an Excepted Risk (see below).

Excepted Risks: This provides a list of items that are not normally covered by the insurance industry, e.g. loss or damage caused by nuclear activity or sonic booms. Fortunately any damage caused by an excepted risk is normally dealt with through Statute, i.e. laws are in place to ensure that any party suffering from an Excepted Risk is able to claim compensation.

Joint Names Policy: This is a policy that is taken out in the name of both the Employer and the Contractor. This is to ensure that both the Employer and Contractor receive the full benefit of the insurance cover. Where an insurance company has made a payment against an insurance policy they are not permitted to try and recover these costs from any person who was insured (or recognised) under the policy. For example, a project has been damaged by fire as a result of the Contractor's negligence and the All Risks Insurance has been taken out by the Employer. The remedial works will be paid for by the insurance company who are prevented from trying to recover their costs from the Contractor because the Contractor has been included in the Joint Names Policy.

The right of insurance companies to recover their costs is closely linked to the legal issue of 'subrogation'. This is a topic that has caused problems in the past in relation to the wording used in some JCT forms of contract, and has led to a number of legal disputes.

(See Petrofina (UK) Ltd & Others v Magnaload & Others, 1984; City of Manchester v Fram Gerrard Ltd, 1974; Ossory Road Ltd v Balfour Beatty Ltd and Others, 1993)

Specified Perils: This identifies the risks that are covered by a Specified Perils policy. Attention is drawn to the fact that 'Excepted Risks' (explained above) are excluded from a Specified Perils policy. In the past there have been disputes about the wording of this definition in particular to the interpretation of what is meant by damage caused through flood or the escape of water from apparatus, etc.

(See William Tomkinson and Sons Ltd v PCC of St Michael in the Hamlet, 1990; Computer and Systems Engineering plc v J Lelliot (Ilford) Ltd & Others, 1990)

Terrorism Cover: This is provided through a Joint Names policy and provides cover for physical loss or damage caused to the Works, Site Materials, and where relevant, existing structures and contents. The loss and damage must have been caused by an act of terrorism. This definition applies to insurance taken out under Insurance Option A, B or C.

Sub-Contractors – Specified Perils Cover under Joint Names All Risks Policies
The purpose of this clause is to give sub-contractors some protection under the insurance cover provided by the All Risks policy taken out by either the Contractor or the Employer.

6.9.1   Depending upon which party takes out the All Risks policy (i.e. the Contractor or the Employer) they must ensure that all sub-contractors are given protection under the policy as required in Schedule 3 paragraphs A.1, A.3, B.1 or C.2. The requirements may be met by one of two means, i.e.:

6.9.1.1   each sub-contractor is to be recognised (i.e. their name may be inserted into the policy) as being insured under the relevant Joint Names Policy, or

6.9.1.2   it may be stated within the policy that the insurer will forego his right of subrogation against any sub-contractor employed upon the Works.

The purpose of clause 6.9.1 is to provide sub-contractors with some of the benefits of the All Risks Insurance. The sub-contractor does not enjoy the full All Risks cover; their cover is limited to that of 'Specified Perils' (see clause 6.8 definitions) in relation to loss and damage to the Works and Site Materials. The sub-contractor enjoys this cover until their sub-contract works are practically complete (this date will be agreed between the Contractor and sub-contractor, see JCT SBCSub/C clause 2.20) or termination of the sub-contractor's employment.

(See Petrofina (UK) Ltd & Others v Magnaload & Others, 1984; British Telecommunications plc v James Thomson & Sons Ltd, 1998)

6.9.2   Where either the Contractor or Employer fails to take out or maintain the All Risks policy, the non-defaulting party may themselves take out the policy (see Schedule 3 paragraph A.2, B.2.1.2 or C.3.1.2). Where the non-defaulting party takes out the policy they must provide the sub-contractor cover as detailed in clause 6.9.1.

Terrorism Cover – Non-Availability – Employer's Options

This section deals with the possibility that the insurance market may decide to withdraw from providing terrorism cover. For example, if the country was to suffer a period of extensive and sustained acts of terrorism, the insurance market might make a commercial decision to stop providing terrorism cover.

6.10.1   This clause relates to the Joint Names policy (for the Works) that may be taken out under Insurance Options A or B, and both the Joint Names policies (i.e. existing structures and the Works) detailed in Option C. If the insurer named in any of these policies notifies either party (Employer or Contractor) of their intention to terminate the Terrorism Cover, then the party receiving the notice must immediately inform the other party of this event. The date, supplied by the insurer, from when the Terrorism Cover ceases is referred to as the 'cessation date'.

6.10.2   Upon receiving a notification as detailed above, and before the cessation date, the Employer must give the Contractor a written notice. In the notice the Employer is to state either:

6.10.2.1   that despite the withdrawal of the Terrorism Cover the Employer still requires the Contractor to carry on and complete the Works, or

6.10.2.2 that the Contractor's employment is to be terminated. The Employer must state a date for this termination. The date cannot be any earlier than the date of the notification and it cannot be any later than the 'cessation date'. If an insurer gives only a short period of notice, the Employer must be advised to consider swiftly whether or not he wishes to continue with the Works. If the Employer delays making a decision and terminates the Contractor's employment after the cessation date, he will be in breach of contract.

6.10.3 Where the Employer does terminate the Contractor's employment under clause 6.10.2.2, then the procedures detailed in clauses 8.12.2 to 8.12.5 are to be followed, with the exception of clause 8.12.3.5. This exception means that the Contractor is not entitled to claim loss and/or damage incurred as a direct result of this termination. Furthermore, as a result of the termination, any other clauses that require further payments to be made, or Retention to be released, are no longer applicable, therefore the Contractor's entitlement to payment is now largely governed by clause 8.12.

6.10.4 Where the Employer does not terminate the Contractor's employment under clause 6.10.2.2, the following procedures are to apply:

6.10.4.1 where the Works or Site Materials have been damaged by terrorism, the Contractor is to restore the damaged works, replace or repair materials and remove any associated debris. The Contractor is expected to carry this out in a timely fashion and to carry on with the completion of the Works;

6.10.4.2 the cost of the restoration work identified above shall be paid to the Contractor as though it were a variation. This means that the Contractor is entitled to be paid for all the restoration works which will be valued in accordance with the procedures detailed in Section 5 of this contract. There are to be no deductions from the payments made under this clause for acts or neglect of the Contractor (or any sub-contractor) which contributed to the loss or damage. For instance, the Contractor may have failed to make the site secure, making it easier for an act of terrorism to take place;

6.10.4.3 this sub-clause relates to a project where Insurance Option C is applicable, i.e. the Works is being carried out in an existing building or is providing an extension to an existing building. In this situation the requirement under clause 6.10.4.1 that the Contractor proceeds with the Works shall not be affected by the fact that the existing building or its contents may have been damaged. However, this sub-clause does not place the Employer under an obligation to reinstate the existing structures.

CDP Professional Indemnity Insurance

This section is relevant only where the Contractor has been required to carry out CDP work. The purpose of this section is to ensure that the Contractor has appropriate professional indemnity insurance in place to provide cover for any claims the Employer may make as a result of negligent or defective design work undertaken by the Contractor or for which the Contractor is responsible (i.e. design work that has been sub-let).

Obligation to Insure

6.11   Where the contract contains an element of CDP work, the Contractor is required to comply with the following:

    6.11.1   once the contract has been entered into, the Contractor, if he has not already done so, must promptly take out a professional indemnity insurance policy. The policy must meet the specific requirements set out in the Contract Particulars, i.e. the level of financial cover required including, where specified, the level of cover for pollution and contamination claims. The details stated in the Contract Particulars are the *minimum* requirements that the Contractor must comply with, consequently a Contractor may take a commercial decision to purchase insurance cover in excess of these requirements.

    6.11.2   the Contractor is to maintain the professional indemnity insurance for the period stated in the Contract Particulars, i.e. 6 years or 12 years from the date of practical completion. This requirement is subject to the Contractor being able to obtain the insurance at 'commercially reasonable rates'. What constitutes commercially reasonable rates is a potential area of debate, but see clause 6.12.

    6.11.3   if requested by the Employer or A/CA, the Contractor must provide documentary proof that the required insurance has been taken out and is being maintained. Any such requests must be reasonable.

Increased Cost and Non-Availability

The insurance market is always subject to change. Situations can sometimes arise where the insurance market decides to limit its exposure to certain risks either by refusing to issue policies or by increasing the cost of the policies to uncommercial levels. This clause identifies the actions to be taken if this situation arises.

6.12   If the professional indemnity insurance is no longer available at commercially reasonable rates the Contractor must immediately notify the Employer. The purpose of this notification is to allow the Contractor and Employer to try and reach an agreement on how this problem may be dealt with to the satisfaction of both parties.

Joint Fire Code – Compliance
Application of Clauses

6.13   If clauses 6.14 to 6.16 are to apply to the contract, this fact must be stated in the Contract Particulars. If the All Risks policy has been taken out under Option A then the details in the Contract Particulars will have to be completed based on information provided by the Contractor (see Footnote [22]).

Compliance with Joint Fire Code

6.14 Both parties, i.e. the Employer and Contractor, must comply with the Joint Fire Code (JFC). The Employer must also ensure that people for whom he is responsible (Employer's Persons) also comply with the JFC. Likewise the Contractor must ensure that people for whom he is responsible (Contractor's Persons) comply with the JFC.

Breach of Joint Fire Code – Remedial Measures

6.15.1 If there is a breach of the JFC the All Risks insurer may notify either party (Contractor or Employer) that remedial measures must be put in place. Whichever party receives this notification must send copies to the other party and the A/CA, and then the following action is to take place:

    6.15.1.1 where the notified remedial measures relate to the Contractor's task of carrying out the Works (e.g. it is a construction, materials or workmanship issue) then the Contractor must comply with the notice. If the insurer has given a specific time by which the work should be carried out, then again the Contractor must comply with this requirement. However, this clause is subject to the provision of the following sub-clause:

    6.15.1.2 if the insurer's notification of remedial measures requires a variation to the Works, then the A/CA must issue an appropriate instruction to allow the Contractor to comply with the notice. However if, as a result of the notice, the Contractor has to carry out all or some of the work before the receipt of an instruction (i.e. in an emergency) he may do so as long as he provides only the minimum amount of work or materials reasonably necessary to comply with the notice. The Contractor must also promptly inform the A/CA of the emergency and the amount of work he is executing in accordance with this clause. Such work is to be treated as though it were a variation (i.e. no variation instruction need be issued but the works will be paid for as though they were covered by a variation instruction). However, if the emergency work relates to CDP then the Contractor is not entitled to receive payment.

6.15.2 If the Contractor fails to commence remedial action within 7 days of receiving a notice, or if he fails to proceed with the remedial measures in a regular and diligent manner, the Employer may engage outside contractors to execute the work. The cost of this work may be recovered from the Contractor by making a deduction from the Contract Sum, i.e. from the next Interim Certificate. This clause applies only to remedial measures relating to the Contractor's work which do not require a variation instruction from the A/CA.

Joint Fire Code – Amendments/Revisions

6.16   If, after the Base Date, there is an amendment or revision to the JFC which requires the Contractor to undertake additional work so as to comply with the amendment or revision, then the cost will be met by the party identified in the Contract Particulars, i.e. Contractor or Employer. If this section of the Contract Particulars is not completed then by default the Contractor is the party responsible for meeting the cost of the amendment or revision.

Section 7: Assignment, Third Party Rights and Collateral Warranties

Assignment
General

7.1 If the Employer or Contractor wishes to assign this contract, or any of the rights contained within the contract, they must obtain the written consent of the other party.

(*Note*: assignment is a fairly complex legal issue but generally when parties enter into a contract it is usually possible for either party to transfer their contractual rights to a third party through the process of an 'assignment', and it is not normally necessary to obtain the permission of the other party. However, as a result of this clause the Employer or Contractor must obtain written permission before assigning their rights.)

Clause 7.1 is slightly modified by the provisions contained within clause 7.2.

7.2 This is an optional clause and it will be necessary to check the Contract Particulars to see whether the clause applies to this contract or not. If this clause does apply it means that if the Employer transfers his freehold or leasehold interest in the Works (i.e. sells or leases the Works) he may, after practical completion, grant or assign to the freeholder or lessee the right to bring proceedings against the Contractor. The proceedings must be brought in the Employer's name and can take the form of arbitration or litigation, depending on what has been stipulated in the contract (see Articles 8 and 9). In the action the third party, through the Employer, has the right to enforce any terms in the contract that are to the benefit of the Employer. A third party cannot attempt to overturn any binding agreements reached between the Contractor and Employer before the date of the grant or assignment. This is true even if the prior agreement has an adverse impact on the grant or assignment.

Similarly, if the Works have been broken down into Sections the Employer may grant or assign the right to bring an action relating to that Section, from the date of the Section Completion Certificate.

Clauses 7A to 7E – Preliminary
References

7.3 Where in these conditions a reference is made to Part 2 (i.e. this refers to Part 2 in the Contract Particulars) it is deemed to include any documents that are referred to in Part 2.

Notices

7.4 Any notice that is referred to in clauses 7A to 7E must be issued in writing, and it must be given to the Contractor by the following means: actual delivery, special or recorded delivery. With reference to the latter two methods of delivery, the rules contained within clause 8.2.3 will determine the date on which the notice is deemed to have been received. Where reference is made to a non-specified collateral warranty (i.e. a collateral warranty other than one of the specified JCT forms) a copy of the collateral warranty must be provided with the notice.

Execution of Collateral Warranties

7.5   Where this contract is executed as a deed (see Attestation pages at the back of the Contract Particulars) then any collateral warranties that are entered into under the provisions of Section 7 of this contract shall also be entered into as a deed. Where this contract is entered into under hand the collateral warranties *may* be executed under hand. This wording would seem to imply that the collateral warranty could also be entered into as a deed, but the period of liability would still be 6 years (see Schedule 5 paragraph 8.1).

Third Party Rights from Contractor
7A Rights for Purchasers and Tenants

7A.1   Where Part 2 of the Contract Particulars specifies that clause 7A applies to a Purchaser or Tenant, their P&T Rights will commence from the date the Contractor receives the appropriate Employer's notice. The notice must identify, by name, the Purchaser or Tenant and their interest in the Works, i.e. a freehold or leasehold interest and whether it applies to the whole or part of the Works.

7A.2   This clause identifies the rights of the Employer and Contractor where clause 7A is operative. The two parties have the right to:

7A.2.1   terminate the Contractor's employment under this contract, whether under a specific term of this contract or by any other means, or they may even terminate the contract itself;

7A.2.2   agree to amend or alter the terms contained within the contract and waive any terms of the contract;

7A.2.3   agree, at their discretion, on how any dispute or problem arising from this contract may be settled.

These rights may be exercised by the Employer or Contractor as they wish; there is no requirement to obtain the consent of the Purchaser or Tenant. However, this is slightly modified by the next clause:

7A.3   Once the Purchaser or Tenant has acquired the P&T Rights neither the Employer nor Contractor may change or alter the express terms of clause 7 A or Part 1 of Schedule 5 without the consent of the Purchaser or Tenant. It can be seen that this clause slightly modifies the provisions contained within clause 7A.2.

7B Rights for Funder

7B.1   Where Part 2 of the Contract Particulars specifies that clause 7B applies to a Funder, the Funder Rights will commence from the date the Contractor receives the required Employer's notice, identifying the Funder.

7B.2   Where a Funder has acquired Funder Rights in accordance with 7B.1, the following provisions will apply:

7B.2.1 there is to be no change or variation to the terms contained within clause 7B or Part 2 of Schedule 5 without first obtaining the written consent of the Funder; and

7B.2.2 neither the Employer nor the Contractor may agree to the termination of this contract. The Contractor's right to terminate his employment can only be exercised after complying with the terms of Schedule 5, Part 2, paragraph 6.1. Likewise the Contractor may not accept the contract as being repudiated without first complying with Schedule 5, Part 2, paragraph 6.2.

However, subject to the above restrictions, the Contractor, in agreement with the Employer, is free to alter, vary or waive the terms of this contract without having to obtain the consent of the Funder. Likewise the Contractor and Employer may agree, at their discretion, on how any dispute or problem arising from this contract may be settled without the Contractor having to obtain the Funder's consent. It is important to be aware that these rights of the Contractor are extinguished once the Funder has issued a notice under Schedule 5, Part 2, paragraph 5 or 6.4. Once this notice has been issued the Contractor is to accept instructions from the Funder to the exclusion of the Employer.

Collateral Warranties
7C Contractor's Warranties – Purchasers and Tenants
Where Part 2 of the Contract Particulars specifies that clause 7C applies to a Purchaser or Tenant, the Employer may give the Contractor a notice requiring him to enter into a collateral warranty with the Purchaser or Tenant. In the notice the Employer must identify the Purchaser or Tenant and what interest they have in the Works. The Contractor is to enter into the JCT standard collateral warranty (CWa/P&T) with the Purchaser or Tenant within 14 days of receiving the Employer's notice. The collateral warranty is to be completed in accordance with the detail provided in Part 2 of the Contract Particulars (i.e. the P&T Rights Particulars).

7D Contractor's Warranty-Funder
Where Part 2 of the Contract Particulars specifies that clause 7D is to apply to a Funder, the Employer may give the Contractor a notice requiring him to enter into a collateral warranty with the Funder. Where there is more than one Funder to the project, the Employer should identify the Funder in the notice. The Contractor is to enter into the JCT standard collateral warranty (CWa/F) with the Funder within 14 days of receiving the Employer's notice. The collateral warranty is to be completed in accordance with the detail provided in Part 2 of the Contract Particulars (i.e. the Funder Rights Particulars).

7E Sub-Contractors' Warranties
Part 2(E) of the Contract Particulars may specify that warranties are required from sub-contractors in favour of a Purchaser, Tenant, Funder or Employer. Where sub-contractor warranties are required, the Contractor is to obtain the warranties as detailed in the

Contract Particulars. The Employer must give the Contractor a notice that identifies the relevant sub-contractor and what type of warranty is required and for whom. Within 21 days of receiving the notice, the Contractor is to obtain the warranties from the sub-contractors. The warranties are to be the JCT standard forms, i.e. SCWa/P&T, SCWa/F or SCWa/E and are to be completed in accordance with the Contract Particulars Part 2(E). A sub-contractor is allowed to suggest amendments to the standard warranties. The amendments must have the approval of the Contractor and Employer; this approval may not be unreasonably delayed or withheld.

Section 8: Termination

General
Meaning of Insolvency

The purpose of clause 8.1 is to define what is meant by the term 'insolvency'.

8.1 For the purposes of this contract the Employer or Contractor is considered to be insolvent if any of the following events arise:

8.1.1 an arrangement, compromise or composition in satisfaction of debts is entered into. This basically refers to a situation where a party has reached agreement with his creditors to clear his debts by paying a lesser amount. For example, creditors may accept a proposal to receive a payment of 20 pence for every pound owed, or they may agree to extend the period for repayment. Creditors will sometimes accept such proposals as they consider it will give them a better return than forcing the party into liquidation (or bankruptcy, if the party is not a limited company);

8.1.2 either the Employer or Contractor passes a resolution, or makes a determination, that they be wound up but they are unable to issue a declaration of solvency.

This situation may arise through a creditor's voluntary liquidation when the directors call a meeting of the shareholders seeking their approval to wind up the company and appoint a liquidator.

Alternatively, if the director's wish to cease trading although the company is solvent, they may wind the company up through a members' voluntary liquidation and make a declaration of solvency – in which case this clause would not apply.

8.1.3 a party has a winding-up order or bankruptcy order made against him. For example, this process may be initiated by a creditor who submits a petition to the court for a winding-up order or a company, through its directors or shareholders, may also present a petition;

8.1.4 a party has an administrator or administrative receiver appointed. Since the Enterprise Act it has become more difficult for a creditor to have an administrative receiver appointed but it has become easier for an administrator to be appointed to oversee a company's affairs;

8.1.5 a party is subject to similar insolvency proceedings as identified above covered by jurisdiction outside the UK;

8.1.6 a party who is a partner within a partnership is the subject of an individual arrangement or any of the above insolvency events. It is important to be aware that this clause applies when any partner within the partnership becomes the subject of an insolvency event, whether or not they are engaged on the project. The reason for this is that all the partners will be equally liable for the insolvency event.

Notices Under Section 8

8.2.1   A Contractor's employment should not be terminated unreasonably or vexatiously. It should be seen to be a reasonable action by the party giving the notice.

8.2.2   The termination will be effective when the party (i.e. either Contractor or Employer) receives the termination notice.

(See JM Hills & Sons Ltd v London Borough of Camden, 1980)

8.2.3   Where a termination notice is given it must be in writing. The notice may be given by actual delivery (i.e. by hand) or by special or recorded delivery. Where the notice is sent by special or recorded delivery it is deemed to have been received by the other party on the second business day from the date of posting, therefore, a notice sent by recorded delivery on Friday 2nd July will be deemed to have been received on Tuesday 6th July. However, either party may challenge the deemed date if they can provide proof the notice was delivered on a different day or possibly not delivered at all.

Other Rights, Reinstatement

8.3.1   The provisions and procedures incorporated within clauses 8.2 to 8.7 do not prevent the Employer from using any other rights or remedies which are available to him, e.g. statute law, contract law or common law. Similarly the Contractor is not limited to the procedures set out in clauses 8.9, 8.10 and 8.12; he is fully entitled to make use of any legal rights he may possess outside the contract conditions.

8.3.2   Regardless of the reason for the termination, the Contractor's employment may be reinstated at any time. The reinstatement would obviously have to be as a result of an agreement between the Employer and Contractor, and may be on the basis of the original contract conditions or the parties may agree new or amended terms.

Termination by Employer
Default by Contractor

8.4.1   Clause 8.4.1 lists the defaults that may entitle the Employer to terminate a Contractor's employment. The list refers to defaults committed by the Contractor before the date of practical completion, i.e.:

8.4.1.1   if, without reasonable cause, the Contractor completely or substantially suspends carrying out the Works or the design of the CDP. Note that there has to be at least a substantial suspension of the Works, not just a minor infringement;

8.4.1.2   if the Contractor fails to proceed *regularly and diligently* with the Works or the design of the CDP, a phrase that has caused a degree of uncertainty in the past – it can be a difficult concept to clearly define and identify (see West Faulkner Associates v London Borough of Newham, 1992);

8.4.1.3 if the Contractor refuses or fails to comply with a proper request of the A/CA to remove defective work or materials, and as a result the Works are materially affected. Note the requirement that the breach must *materially* affect the Works; this clause could not be used for a minor irregularity;

8.4.1.4 if the Contractor assigns the contract without the Employer's agreement (see clause 7.1), or sub-lets part of the Works without the A/CA's consent or sub-lets the design for CDP work without the Employer's consent (see clause 3.7);

8.4.1.5 the Contractor fails to comply with the CDM Regulations (see clause 3.25).

If the Contractor commits one or more of the above, the A/CA *may* give him a notice identifying the default(s), i.e. a specified default. Note that the A/CA is not obliged to issue a default notice; it is at his discretion whether or not he does so. Any notice issued must comply with clause 8.2.3.

8.4.2 If the Contractor continues with a specified default for 14 days after he receives a default notice, then within 10 days from the expiry of the warning notice (i.e. from day 15 up until day 24) the Employer may send the Contractor a written notice of termination. The termination will be effective on the day the Contractor receives the termination notice. For instance, a Contractor receives a default notice from the A/CA on 31st July, and by 14th August it is evident the Contractor has taken no steps to remedy the default therefore the Employer may issue the Contractor with a termination notice at any time from 15th to 24th August. Again it should be noted that an Employer is not obliged to issue a termination notice; it is an option that he may use if he wishes – see next clause.

8.4.3 If the Contractor ceases the default within 14 days of the default notice, or the Employer fails to give a notice of termination within the 10-day period allowed, the Employer's right of termination lapses. However, if the Contractor then repeats a specified default the Employer's right of termination is resurrected and he is entitled to give a termination notice immediately upon the date of the reoccurrence or within a *reasonable* time of the repeated default. It is important to note that, where there is a repeated default, the Employer is no longer limited to giving the Contractor a termination notice within a 10-day period. Any termination notice may be given within a reasonable time of the repeated default. What is deemed to be a reasonable time will depend upon individual circumstances.

Insolvency of Contractor

8.5.1 If the Contractor is insolvent (see definitions of insolvency in clause 8.1) the Employer may terminate the Contractor's employment at any time by issuing a termination notice.

8.5.2   The Contractor is under an obligation to immediately notify the Employer if he proposes to commence any of the proceedings identified in clause 8.1, or if he becomes subject to any proceedings under clause 8.1. The Contractor's notice must be in writing. The purpose of sub-clause 8.5.2 is to ensure that the Employer is given prompt notice of any likely insolvency event.

8.5.3   From the date the Contractor becomes insolvent (as defined in clause 8.1), and regardless of whether or not the Employer has terminated the Contractor's employment, certain procedures are to apply, i.e.:

8.5.3.1   the provisions of clauses 8.7.4, 8.7.5 and 8.8 shall apply as if a termination notice had been given. Any contract conditions that require further payments to be made or Retention monies to be released are no longer applicable;

8.5.3.2   the Contractor is no longer required to carry out and complete the Works (clause 2.1) or carry out the design for CDP work (clause 2.2). These obligations under clauses 2.1 and 2.2 are suspended, i.e. held in abeyance until further decisions are made regarding the insolvency;

8.5.3.3   the Employer is entitled, without interference from the Contractor, to take reasonable steps to protect and secure the site, work and Site Materials. Following insolvency the site may be virtually abandoned by the Contractor and become subject to vandalism, theft and the 'recovery' of materials by potential creditors. This clause entitles the Employer to secure the site and thereby prevent such actions.

Corruption

8.6   The Employer may terminate the Contractor's employment if the Contractor (or any person employed by him or acting for him) commits an offence under the Prevention of Corruption Acts 1889 to 1916. This may relate to an offence committed on this contract or any other contract between the Employer and Contractor. Furthermore, if the Employer is a Local Authority, the Contractor's employment may also be terminated where the Contractor has given a fee or reward in contravention of the Local Government Act 1972.

Consequences of Termination Under Clauses 8.4 to 8.6

8.7   If the Contractor's employment is terminated under clauses 8.4 to 8.6, then the following procedures are to apply:

8.7.1   the Employer may engage and pay other contractors to carry out and complete the Works and to make good any defects in the Contractor's workmanship and materials. To allow the Works to be continued the Employer (and any contractors that may have been engaged by the Employer) has the right to take possession of the site and the Works and use any temporary buildings, plant, tools, equipment and Site Materials owned by the Contractor. If these items are not owned by the Contractor then the Employer must obtain the owner's consent before they may be used. The Employer is also entitled to engage others to carry out and complete the design work for any CDP work.

8.7.2   the obligations placed upon the Contractor are as follows:

8.7.2.1   the Contractor is to remove his temporary buildings, plant, tools materials, etc. and at the same time arrange for the removal of temporary buildings, etc. not belonging to him. These items are to be removed only after receipt of a written request from the A/CA;

8.7.2.2   where the project contains CDP work, the Contractor is to provide the Employer with 2 copies of all Contractor's Design Documents that have been prepared to date. The Contractor must comply with this requirement even if he has previously provided the Employer with all or some of these documents. The documents are supplied to the Employer free of charge;

8.7.2.3   within 14 days of the termination the Employer or A/CA can require the Contractor to assign, to the Employer, the benefits of any contracts for the supply of materials and goods and/or contracts for the execution of Works. This is to be done free of charge to the Employer and only where the sub-contract or contract of sale permits such an assignment. The JCT through Footnote [59] advises that the Employer may not be able to exercise this right if the Contractor's employment was terminated because of an insolvency event listed in clause 8.1, i.e. the appointment of an administrative receiver or administrator, a winding-up order being made, the passing of a resolution for a voluntary winding-up, a proposal for a voluntary arrangement, etc.

8.7.4   Once the Works have been completed, and any defective work of the defaulting Contractor has been dealt with, a financial account is to be prepared. It is to be prepared within a reasonable time and may be submitted as a statement from the Employer or as an A/CA's certificate.

The following sub-clauses set out the structure of the financial account and determine whether the defaulting Contractor is entitled to any further monies or, the more likely case, whether the Contractor owes money to the Employer. The statement or certificate is to identify the following:

8.7.4.1   the amount of expenses properly incurred by the Employer as a result of the termination of the Contractor's employment. The expenses may include any costs incurred through the operation of clause 8.5.3.3 (securing the site) and clause 8.7.1 (completing the Works) plus any direct loss and/or damage caused to the Employer for which the Contractor is liable, whether or not it was caused by the termination;

8.7.4.2   the amount of payments that have been made to the defaulting Contractor;

8.7.4.3   the total amount that would have been payable for the Works if the original Contractor had completed the Works in accordance with the contract conditions. To arrive at this figure the Employer or A/CA would have to request the QS to prepare a notional final account for the defaulting Contractor.

8.7.5   The amounts from clauses 8.7.4.1 and 8.7.4.2 are totalled, and if that figure exceeds the notional final account from clause 8.7.4.3, the difference will be a debt due from the Contractor to the Employer. If the notional final account exceeds the total amount from clauses 8.7.4.1 and 8.7.4.2 (this rarely occurs) then the difference is a debt due from the Employer to the Contractor.

Employer's Decision Not to Complete the Works

Following the termination of a contractor's employment it would be assumed that most Employers would look for a suitable method by which their project may be completed. However, there may be occasions when the Employer no longer wishes to continue with the project (see Tern Construction v RBS Garages, 1992). The following clauses detail the procedures to be followed when the employer decides not to complete the works.

8.8.1   If, within 6 months from the date of termination of the Contractor's employment, the Employer decides not to continue with the Works he must immediately give the Contractor a written notification. Within a reasonable time from sending the written notification the Employer *must* send the Contractor a statement.

If the Employer has made no arrangements for the work to be completed within 6 months of the termination but has not sent the Contractor a notification to that effect, then the Employer must send the Contractor a statement.

The statement must provide the following information:

8.8.1.1   the total value of work properly executed at the date of termination (or, where relevant, at the date when the Contractor became insolvent) calculated in accordance with the contract conditions, plus any other payments which the Contractor is entitled to receive under the contract conditions;

8.8.1.2   the amount of any expense properly incurred by the Employer as a result of the termination plus any direct loss and/or damage for which the Contractor is liable to the Employer, regardless of whether it was caused by the termination or not.

8.8.2   By comparing the amounts calculated under clauses 8.8.1.1 and 8.8.1.2, and taking into account the amounts that have been previously paid to the Contractor, it is possible to ascertain whether there is a debt due from the Contractor to the Employer, or vice versa.

As an example, a Contractor has received £3,453,200 through Interim Certificates to date; the value of work properly executed at the date of termination has been assessed as £3,597,653 and other payments which the Contractor is entitled to are £30,564; the Employer's expenses and direct loss and/or damage have been calculated to be £44,397. The Contractor is entitled to receive the payment detailed below:

Statement of Account	£	£
Value of work properly executed		3,597,653.00
Other payments		30,564.00
Sub-total		3,628,217.00
Less:		
Previous payments	3,453,200.00	
Employer's expenses, etc.	44,397.00	
Sub-total	3,497,597.00	3,497,597.00
Amount due to Contractor		**130,620.00**

Termination by Contractor

Default by Employer

8.9.1 This clause identifies a number of Employer defaults that may lead to the Contractor terminating his employment, e.g. if the Employer:

8.9.1.1 fails to pay the Contractor, by the final due date, monies properly due under a certificate, including any associated VAT. For example, the Employer fails to pay the full amount certified in an Interim Certificate within 14 days from its date of issue and has failed to issue a payment or withholding notice explaining the discrepancy;

8.9.1.2 interferes with the issue of a certificate that is due under the contract conditions, or obstructs the issue of a certificate; for example, if an Employer were to instruct the A/CA not to issue a Practical Completion Certificate even though the Works were complete;

8.9.1.3 fails to comply with clause 7.1, i.e. the Employer assigns the contract without the Contractor's written consent;

8.9.1.4 fails to comply with the CDM requirements set out in clause 3.25.

If any of the above events occur the Contractor *may* give the Employer a notice identifying the default(s). *Note*: the Contractor is not obliged to give the Employer a notice, it is at the Contractor's discretion.

8.9.2 This clause refers to a situation where, before practical completion, the Works or a substantial part of the Works is suspended because of an Employer or A/CA related default or action, and the period of suspension exceeds the time allowed for in the Contract Particulars (any time period may be inserted into the Contract Particulars; the default period is two months). The period of suspension has to be continuous, for example, work suspended from

March 1 to April 28th and again from May 14th to May 31st would have to be treated as two separate periods of suspension, and neither of these periods would, individually, exceed the default period of two months. The acts or defaults relevant to the operation of this clause are as follows:

8.9.2.1   the A/CA issues instructions under:

clause 2.15 – divergence between Contract Documents;

clause 13.4 – variations;

clause 13.5 – postponement of any of the Works;

8.9.2.2   any impediment, prevention or default by the Employer, the A/CA, the QS or any of the Employer's Persons (see Definition clause 1.1). The impediment, etc. may be the result of a specific act or the failure to do something.

There is a proviso that this clause will not take effect if any of the above acts or defaults have been caused by the negligence or default of the Contractor (or any Contractor's Persons). Also, this clause cannot be used where the problems have arisen because of a default or negligence of a Statutory Undertaker, resulting in the issuing of instructions under clause 8.11.1.2. Otherwise, in response to any of the above events, the Contractor may give the Employer a notice specifying the event(s). Note that this is optional for the Contractor; he is not obliged to issue the Employer with a notice of a specified suspension event. A notice could be given to the Employer if the Contractor wanted to terminate, or threaten to terminate his employment, or the Contractor may use the notice as leverage to bring the suspension of the Works to an end.

8.9.3   If the Employer continues with a specified default, or a specified suspension event continues for 14 days after receipt of the Contractor's notice; then on or within 10 days from the end of this 14-day period the Contractor may, by issuing a further notice, terminate his employment. The Contractor's employment is terminated on the day the Employer receives the second notice.

8.9.4   If the Employer ends a specified default, or a specified suspension event ceases within the 14-day period, or the Contractor does not issue the termination notice as above, the Contractor loses his right of termination under clause 8.9.3. However, this right of termination is revived if:

8.9.4.1   the Employer subsequently repeats a specified default, or

8.9.4.2   a specified suspension event occurs again for any period of time whereby the regular progress of the Works is, or is likely to be, affected. Note that in this instance there is no period of time specified for the suspension, but there is a criterion that the suspension must have an impact on the progress of the Works.

In response to a repeated default or suspension, the Contractor may give a notice of termination immediately upon the repetition of the default, etc., or within a reasonable time of the repetition. The termination will be effective on the day the Employer receives the notification.

Insolvency of Employer

8.10.1   This clause contains the procedures to be followed when the Employer has insolvency problems; where an Employer is insolvent (see clause 8.1), the Contractor may terminate his employment by issuing an appropriate notice to the Employer.

8.10.2   The Employer is under an obligation to immediately notify the Contractor if he proposes to commence any of the proceedings identified in clause 8.1 or if he becomes subject to any proceedings under clause 8.1. The Employer's notice must be in writing. The purpose of this clause is to ensure that the Contractor is given prompt notice of the Employer's insolvency or impending insolvency.

8.10.3   From the date of the Employer's insolvency, the Contractor's obligation to carry out and complete the Works (clause 2.1) and to carry out the design for CDP work (clause 2.2) is suspended.

Termination by Either Party

8.11.1   This clause relates to a situation where the Works have not yet reached practical completion and the execution of the uncompleted Works, or a substantial part thereof, has been suspended for any of the reasons given below and for a continuous period of time in excess of the 'period of suspension' stated in the Contract Particulars (the default period is 2 months, but it is necessary to check the contract to see if an alternative time period has been inserted). This procedure may be triggered off by any of the following events:

8.11.1.1   force majeure; the JCT does not provide a definition of this term but it is generally thought to refer to exceptional matters that are beyond the control of the parties;

8.11.1.2   instructions issued by the A/CA under clause 2.15 (instructions regarding discrepancies or divergences between the Contract Documents), clause 3.14 (variation instructions) or clause 3.15 (postponement of any work) where issued as a result of a default by a Statutory Undertaker;

8.11.1.3   where loss or damaged has occurred to the Works as the result of a 'Specified Perils' event;

8.11.1.4   civil commotion, an act of terrorism or the threat of a terrorist act. Also, the action taken by an authority in dealing with the civil commotion, terrorist act or threat; or

8.11.1.5   the Government's using its statutory powers in a way that directly affects the carrying out of the Works.

Consequently, if any of the above events arises leading to the suspension of all or part of the Works, either party may give the other a written notice. This notice may only be given once the period of suspension (see clause 8.11.1) has been exceeded. The notice must state that the Contractor's employment will be terminated unless the suspension is lifted within 7 days from the date the party received the written notice. If the suspension is not lifted within the 7-day period, then the party issuing the first notice may now issue a further notice terminating the Contractor's employment. The Contractor's right of determination under sub-clause 8.11.1.3 is slightly modified by the following clause.

8.11.2   A Contractor may not give a notice of determination under clause 8.11.1.3 (suspension caused by Specified Perils) where the loss or damage has been caused by his negligence or that of any 'Contractor's Persons'.

Consequences of Termination Under Clauses 8.9 to 8.11, etc.

8.12   Where the Contractor's employment has been terminated under clauses 8.9 to 8.11, 6.10.2.2 (i.e. withdrawal of Terrorism Cover) or Schedule 3, paragraph C.4.4 (i.e. damage to the Works covered by an All Risks policy) the following procedures are to apply:

8.12.1   upon the termination of the Contractor's employment, clause 8.12 is to become operative, and other clauses within the contract requiring further payments to be made to the Contractor, or Retention to be released, are no longer applicable;

8.12.2   the Contractor's obligations are as follows:

8.12.2.1   the Contractor and his sub-contractors are to remove, with reasonable promptness, any temporary buildings, plant, tools and equipment belonging to them. They are also to remove all goods, materials and Site Materials (even though some of the Site Materials may have been paid for by the Employer, but see clause 8.12.5);

8.12.2.2   where the project includes an element of CDP work, the Contractor must provide the Employer with 2 copies of the 'as-built drawings' referred to in clause 2.40, but only those drawings that the Contractor had prepared prior to termination.

8.12.3   Where the Contractor's employment is terminated under clauses 8.9 or 8.10 (default or insolvency of the Employer) he is to prepare a financial account as soon as reasonably practical. Where the termination arises under clause 8.11 (termination by either party), clause 6.10.2.2 (withdrawal of Terrorism Cover) or Schedule 3, paragraph C.4.4 (damage to the Works covered by an All Risks policy) the method of preparing the financial account is decided by the Employer. The Employer may require the Contractor to prepare the account or, alternatively may ask the Contractor to provide all necessary documents to enable the Employer to prepare the account. If the latter option is chosen, the Contractor must supply the information within 2 months from the date of termination; the Employer is expected to prepare the account promptly but at the very latest within 3 months of receiving the documents from the Contractor. The content of the account is determined by clauses 8.12.3.1 to 8.12.3.4 and 8.12.3.5 where applicable, and must include the following:

8.12.3.1 the total value of work properly executed at the date of termination (which is to be calculated in accordance with the contract conditions) as if no termination had taken place, plus any other payments which the Contractor is entitled to receive under the contract conditions;

8.12.3.2 any monies which have been assessed for direct loss and/or expense under clauses 3.24 and 4.23. This applies to claims where the value has been assessed before the date of termination and those assessed after the date of termination;

8.12.3.3 any reasonable costs that have been incurred in the removal of temporary buildings, etc. as detailed in clause 8.12.2.1;

8.12.3.4 the cost of Site Materials and any other goods and materials (e.g. this could be items for temporary works or site set-up) that have been properly ordered for the Works and paid for by the Contractor or for which he is legally bound to pay;

8.12.3.5 any loss and/or damage caused to the Contractor and which is directly attributable to the termination (but see clause 8.12.4);

8.12.4 the account may only incorporate the loss and/or damage referred to in clause 8.12.3.5 in certain circumstances, as detailed below:

8.12.4.1 where the termination occurred as a result of clause 8.9 or 8.10; or

8.12.4.2 where the termination occurred as a result of clause 8.11.1.3, and the loss and damage was caused by a specified peril resulting from the Employer's (or Employer's Persons) default or negligence;

8.12.5 once the Employer has submitted the account to the Contractor (or *vice versa*) the Employer must, within 28 days, pay the amount in full (i.e. no Retention may be deducted), less any amounts previously paid to the Contractor under the contract conditions. Where the account has included payment for materials, etc. under clause 8.12.3.4, it is a condition that these materials, etc. will become the property of the Employer.

Section 9: Settlement of Disputes

Mediation

9.1   This clause is merely a recommendation to consider mediation as a means of resolving a dispute, i.e. an attempt to resolve differences through the introduction of a third party who will endeavour to help the parties arrive at a settlement. Mediation among other forms of 'Alternative Dispute Resolutions' is considered by many to be a better method of resolving commercial disputes than litigation, arbitration or adjudication. However, it is not possible for the JCT to make the use of mediation compulsory as this would fall foul of the HGCRA (see R G Carter Limited v Edmund Nuttal Ltd, 2000; John Mowlem & Company plc v Hydra-Tighe Ltd, 2000).

The JCT do not provide any mediation guidelines to complement the 2005 edition of the contract, although they did publish a Practice Note in 1995 to provide advice and documentation for setting up a mediation agreement.[a]

Adjudication

9.2   Where a dispute or difference arises relating to this contract, either party may refer the matter to adjudication. The JCT has decided not to prepare its own adjudication procedures and prefers to rely on the procedures set out in the 'Scheme'. The Scheme for Construction Contracts (England and Wales) Regulations 1998 Part 1 contains a set of adjudication rules prepared by the Parliamentary legislatorial body, and it is these rules the JCT has decided will apply to this contract. However, the JCT has made a few minor amendments to the Scheme as identified below:

9.2.1   The person chosen to act as adjudicator may be named in the Contract Particulars (see the Scheme Part1, 2.1.a); it is not compulsory for the adjudicator to be named. Secondly a 'nominating body' must be identified in the Contract Particulars, this is a specific requirement and this information must be provided in the Contract Particulars; this differs from the Scheme where the naming of a nominating body is optional.

9.2.2   This clause relates to any dispute that may arise over the operation of clause 3.18.4, e.g. a dispute as to whether or not an instruction to carry out further tests to identify possible defective work is reasonable. In such an event the following procedures are to apply:

9.2.2.1   the adjudicator chosen to deal with a dispute under clause 3.18.4 should, wherever practicable, have the appropriate expertise and experience in relation to that instruction or dispute. For example, if the disputed work related to complex mechanical services it would be advisable to seek an adjudicator with such a background;

9.2.2.2   if the adjudicator does not have the required expertise and experience, he is required to appoint a suitable independent expert whose duty it will be to advise the adjudicator and provide him with a written report setting out whether or not the instruction issued under clause 3.18.4 was reasonable in the circumstances.

---

[a] Practice Note 28, Mediation on a Building Contract or Sub-Contract Dispute, RIBA Publications, 1995.

Arbitration
Conduct of Arbitration

9.3 If a dispute is referred to arbitration in accordance with Article 8, the conduct of the arbitration must comply with the JCT 2005 edition of the Construction Industry Model Arbitration Rules (CIMAR) that were current at the Base Date. If there has been a change in the Rules since the Base Date, the parties may agree to adopt the new rules, in which case they must inform the Arbitrator by issuing a joint notice in writing.

A copy of the JCT 2005 edition of the Construction Industry Model Arbitration Rules may be downloaded from the JCT website.

Notices of Reference to Arbitration

Note: where, in the following clauses, a reference is made to a 'Rule' it is referring to the Rules contained within the JCT 2005 edition of the CIMAR.

9.4.1 If, in accordance with Article 8, a party wishes to refer a dispute to arbitration, they must inform the other party by serving them with a written notice. This written notice must comply with Rule 2.1; i.e. it must identify the dispute and require the other party to agree to the appointment of an arbitrator. The parties must agree, within 14 days from the issue of the arbitration notice, on who is to act as the Arbitrator (this period may be extended by the agreement of both parties). If an agreement is not reached within the set time then either party may (in accordance with Rule 2.3) request that an Arbitrator be appointed by the person identified in the Contract Particulars (e.g. President or Vice-President of the Royal Institution of Chartered Surveyors).

9.4.2 If there are a number of related arbitral proceedings arising from the Works, and they are being dealt with under separate arbitration agreements, then Rules 2.6, 2.7 and 2.8 will apply.

Explanation of Rules 2.6 to 2.8

Where a number of arbitral proceedings, on the same project, arise under separate arbitration agreements, the person appointing the arbitrator must duly consider whether the same arbitrator should be appointed for all or some of the proceedings or whether a different arbitrator should be appointed. It is expected that the same arbitrator would be appointed unless there are good grounds for not doing so (Rule 2.6).

Where different persons have the task of appointing an arbitrator in accordance with Rule 2.6, they must each consult with all the others as part of their 'due consideration'. Where an arbitrator has already been appointed for an arbitral proceeding, due consideration should be given to appointing the same arbitrator to the other proceedings (Rule 2.7).

Where two or more parties are required to appoint an arbitrator, their obligation to comply with Rules 2.6 and 2.7 will be discharged if they make an arrangement for another person (or organisation) to make the appointment for disputes covered by Rule 2.6 (Rule 2.8).

9.4.3   After an Arbitrator has been appointed it is possible for either party to refer a further dispute to that Arbitrator. The new dispute must come within the scope of Article 8 and the further notice of arbitration must be given to the other party and the Arbitrator. In such a situation Rule 3.3 shall apply.

Explanation of Rule 3.3.

Further disputes may be referred to the Arbitrator by either party. However, if the non-referring party does not consent to this referral, the Arbitrator may, as he considers appropriate, take the following action:

i)   decide that the dispute be referred and consolidated with the same arbitral proceedings, or

ii)   decide that the dispute should not be referred and consolidated with the same arbitral proceedings. In which case the fresh dispute will be covered by Rules 2.3 and 2.4.

Powers of Arbitrator

9.5   Subject to Article 8 and clause 1.10 (Effect of the Final Certificate) the Arbitrator has the power to rectify the contract so that it truly represents what the parties originally intended (i.e. may remove unintentional errors from within the documentation). He has the authority to have necessary measurements and valuations carried out to help him determine the parties' rights. He may calculate and award sums of money which should have been included in any certificate and, likewise, any money which ought to have been the subject of a certificate. He may review and alter any certificate, opinion, decision, requirement or notice. He is to determine all matters under dispute that have been referred to him in the manner as set out above, he is not bound by any previously issued certificates, opinions, decisions, requirements or notices.

This clause identifies the specific powers possessed by the Arbitrator in addition to his general powers. The Arbitrator is given very broad powers and may alter any certificate, etc. issued by the A/CA as long as it does not conflict with clause 1.10. For example, if a disagreement over the award of an extension of time is referred to the Arbitrator more than 28 days after the issue of the Final Certificate, the Arbitrator could not review the A/CA's award as the Final Certificate is now conclusive evidence that all extensions of time have been properly awarded.

Effect of Award

9.6   The Arbitrator's award is final and binding on the parties, with the exception that a question of law may be subsequently referred to the courts in accordance with the Arbitration Act (see clause 9.7). If this is the case then the referring party must give notice to the other party and to the Arbitrator.

Appeal – Questions of Law

9.7   It is a condition of this contract that the parties have agreed that a question of law may be referred to the courts in accordance with Section 45(2)(a) and Section

69(2)(a) of the Arbitration Act 1996. Either party may make the application but must give notice to the other party and to the Arbitrator. As a consequence a party may:

9.7.1 apply to the courts to determine any question of law arising in the course of the reference; and

9.7.2 appeal to the courts on any question of law that has arisen from the award made by the Arbitrator under this arbitration agreement.

Arbitration Act 1996

9.8 Any arbitration proceeding will be carried out in accordance with the Arbitration Act 1996 (including any subsequent amendments) regardless of where the arbitration (or part of the arbitration) is held.

Schedules

Schedule 1: Contractor's Design Submission Procedure

Schedule 1 is only relevant where contractor's design portion work has been included in the project. This schedule should be read in conjunction with clause 2.9.3, where reference is made to the contractor's design documents (see Definitions clause 1.1 for an explanation of these documents). The general requirement of clause 2.9.3 is that the Contractor must provide the A/CA with further drawings and information (i.e. the contractor's design documents) in relation to the CDP work. This information will normally be provided in accordance with the following schedule, unless an alternative procedure is stated in the contract documents; for example, the employer's requirements may set out procedures different from schedule 1 – in which case the employer's requirements will take precedence. The procedure in Schedule 1 gives the A/CA the opportunity to review the Contractor's Design Documents (CDD) and allows him to pass comment on whether or not he considers the documents to be in compliance with the contract. It is important to note that the A/CA is here merely reviewing and commenting on the contractor's documents and this procedure does nothing to remove or diminish the Contractor's responsibility to ensure that his CDP work is in compliance with this contract. The Contractor's design submission procedure is as follows:

1. This requires the Contractor to prepare the contractor's design documents and provide the A/CA with two copies of each document. Where the employer's requirements or contractor's proposals specify the format for some or all of these documents then the Contractor must ensure that the documents do comply with that format. The Contractor must submit the documents to the A/CA in sufficient time to allow the A/CA to make any comments (i.e. whether the documents are in accordance with the contract – and if not why not) and for those comments to be incorporated into the design before the relevant CDP work is carried out.

2. Within 14 days of receiving two copies of any contractor's design document the A/CA must return one copy to the Contractor. However, a later timescale may apply where an alternative approach has been stated in the contract documents (i.e. the employer's requirements may include specific dates or time periods for submission). In this instance the A/CA must return one copy of the contractor's design documents within 14 days from either the date stated in the contract documents or the end of the period for submission of the documents as stated in the contract documents.

    Each returned copy must be marked up by the A/CA with the letter 'A', 'B' or 'C'. The A/CA should mark a document with a 'B' or 'C' only where he considers the document does not comply with the contract.

3. If the A/CA fails to return a copy of the contractor's design document within the timescale identified in Paragraph 2, then it is deemed to have been marked with the letter 'A'.

4. Where the A/CA has marked a contractor's design document with a letter 'B' or 'C' (indicating that the document does not comply with the contract) he must

identify, through a written comment, why he considers the document does not comply with the contract.

5. This paragraph explains what actions a Contractor may take once he receives a copy of his Contractor's design document from the A/CA:

    5.1 if the document has been marked with the letter 'A' then the Contractor may proceed with that part of the CDP Works strictly in accordance with that document;

    5.2 if the document has been marked with a letter 'B', the Contractor *may* proceed with that part of the CDP Works in accordance with the document as long as he has incorporated the comments of the A/CA into the document. A copy of the amended document must be sent promptly to the A/CA;

    5.3 if the document has been marked with the letter 'C' the Contractor is to consider the comments made by the A/CA and then take one of two courses of action. The Contractor may either amend the document and resubmit it to the A/CA for comment, all in accordance with the procedures set out in paragraph 1, or alternatively, he may challenge the comments of the A/CA through the procedures set out in paragraph 7 below.

6. Where the A/CA has returned a copy of a Contractor's Design Document marked with the letter 'C' the Contractor is not to proceed with any CDP work contained within that document. The Employer has no liability to pay the Contractor for any work executed in accordance with a Contractor's Design Document that has been marked with a 'C' or where the work was not carried out in accordance with the Contractor's Design Documents marked 'B' or 'C'.

7. If the Contractor disagrees with the comments of the A/CA (i.e. where the Contractor's Design Documents have been returned, marked with a 'B' or 'C') and he considers the documents to be in accordance with the contract, then he may challenge the comments. To do this the Contractor must write and notify the A/CA that, in his opinion, compliance with the A/CA's comments would give rise to a variation. In the notification the Contractor must provide a statement setting out his reasons why he considers the comments would give rise to a variation (e.g. the Contractor would have to explain why he thinks the document originally supplied to the A/CA did comply with the contract). The Contractor's written notification must be sent to the A/CA within 7 days from when the Contractor received the comments. On receiving the Contractor's notification, the A/CA has 7 days to respond. In his response the A/CA may either confirm his comments, in which case the Contractor must amend the document in compliance with the comments and resubmit the amended document or alternatively the A/CA may withdraw his comments.

8. The parties need to be aware of the following details when using the Contractor's design submission procedure:

    8.1 Where, in accordance with paragraph 7, the A/CA confirms or withdraws a comment, this does not mean that the Contractor's Design Document, or amended document, has been accepted by the Employer or A/CA as being in accordance with the contract. Also, where the Contractor complies with an A/CA's confirmed comment this does not signify that a variation has been created.

8.2 Where the A/CA has made a written comment on a Contractor's design document, and the Contractor has not challenged it under paragraph 7 the comment will not be treated as giving rise to a variation.

8.3 This paragraph confirms that it is the Contractor's responsibility to ensure that the Contractor's Design Documents and CDP work are in accordance with this contract. This obligation is not removed or reduced by the fact that the Contractor has complied with the submission procedure in Schedule 1, or that the A/CA has commented on the documents submitted.

Schedule 2: Quotation

Submission of Quotation

Schedule 2 is only relevant where the A/CA has issued a variation instruction requesting the Contractor to provide a Schedule 2 Quotation, therefore this Schedule should be read in conjunction with clause 5.3. The purpose of a Schedule 2 Quotation is to allow the Contractor to pre-price a variation before an instruction is issued for the work to proceed.

1.1 A Schedule 2 instruction should contain enough information to allow the Contractor to provide a Schedule 2 Quotation in accordance with paragraph 2. Footnote [62] provides advice on the format which should be used when providing the information; the general advice is that any information should be provided in a format similar to that of the documents provided at the tender stage. Where the information is provided in the format of an addendum bill, it is important to be aware of the requirements of clauses 2.13 and 2.14. These clauses set out the measurement rules which should be followed in the preparation of the addendum bill and the means of dealing with any errors in the addendum bill. If the Contractor reasonably considers the information provided with the A/CA instruction to be inadequate, he should notify the A/CA. The Contractor's notification must be given within 7 days from when he received the instruction. Upon receipt of the notification, the A/CA should provide the necessary information.

1.2 The Contractor is to submit his Schedule 2 Quotation to the QS within 21 days from whichever is the later of the following events:

1.2.1 the date the Contractor received the instruction, or

1.2.2 the date the Contractor received sufficient information as detailed in paragraph 1.1.

1.3 The Schedule 2 Quotation remains open for acceptance, by the Employer, for 7 days from receipt by the QS.

1.4 No work is to be executed against a Schedule 2 Quotation until the Contractor is in receipt of a 'Confirmed Acceptance' issued by the A/CA in accordance with paragraph 3.2.

Content of Quotation

2. This paragraph sets out the information which a Contractor must provide when submitting his Schedule 2 Quotation, i.e.:

2.1 The cost of carrying out the variation (including any allowance for the effect the variation may have on *other* work). The Contractor is to provide a breakdown of the quotation using rates and prices from the Contract Bills where appropriate, and must make due allowance for any preliminary items which may be affected. Any associated direct loss and/or expense should not be included in this calculation as it should be dealt with separately under paragraph 2.3.

2.2 Any adjustment of time required by the Contractor to the Works or any relevant Section, as a consequence of carrying out the variation work. As a result the Contractor may need to ask for additional time, or on some occasions he may require less time, which would result in an earlier contract Completion Date. An important proviso is that the time adjustment should not have been previously allowed by the A/CA under clause 2.28 or claimed by the Contractor in a previous Schedule 2 Quotation.

2.3 Any amount to be paid in lieu of a clause 4.23 claim for direct loss and/or expense. Again it is a requirement that this amount should not have been previously dealt with in a clause 4.23 claim nor included in a previous Schedule 2 Quotation.

2.4 A fair and reasonable amount to cover the Contractor's cost of preparing the Schedule 2 Quotation.

2.5 Over and above these standard requirements the A/CA may, through the instruction, require the Contractor to provide further indicative information in the form of statements, i.e.:

2.5.1 the additional resources which the Contractor may need to carry out the variation, and

2.5.2 the method of carrying out the variation, e.g. a Contractor's method statement.

The Contractor is required to provide reasonably detailed information to support each part of his Schedule 2 Quotation to enable the Employer or his representative to properly evaluate the quotation.

Acceptance of the Quotation

3.1 If the Employer decides to accept the Schedule 2 Quotation, he must notify the Contractor, in writing, within the timescale set down in clause 1.3.

3.2 Where the Employer accepts a Schedule 2 Quotation, the A/CA must provide the Contractor with an immediate written notice of acceptance, i.e. a 'Confirmed Acceptance'. The A/CA in his confirmation is to inform the Contractor of the following:

3.2.1 that the Contractor is to proceed with the variation,

3.2.2 the amount to be paid for the variation including adjustments for any direct loss and/or expense and the cost of preparing the quotation, and

3.2.3 the time adjustment required by the Contractor for the Works or Section, and the revised Completion Date. Advice is also provided to the effect that it is possible for a revised Completion Date to be set that is earlier than the Date for Completion given in the Contract Particulars. This is the only occasion when the A/CA may set a revised Completion Date earlier than the original contract Completion Date.

Quotation Not Accepted

4. If the Employer does not accept the Schedule 2 Quotation within the timescale provided for in paragraph 1.3 then, as soon as the time period has expired,

   4.1 the A/CA shall instruct the Contractor to carry out the work as a variation which will be valued under the Valuation Rules (clauses 5.6 to 5.10). The significance of this is that the work will be valued by the QS and the procedure for the Employer and Contractor to agree the value of the variation under clause 5.2.1 is precluded, or

   4.2 the A/CA shall instruct the Contractor that the variation is not to be carried out.

Costs of Quotation

5. If the Employer does not accept a Schedule 2 Quotation he is still obliged to reimburse the Contractor a 'fair and reasonable' amount for the cost of preparing the quotation – as long as the Contractor had submitted a 'genuine' quotation in the first instance. The mere fact that the Employer has rejected a Schedule 2 Quotation does not imply that it was not prepared on a fair and reasonable basis.

Restriction on Use of Quotation

6. If the A/CA does not issue a Confirmed Acceptance under paragraph 3.2, then neither the Employer nor the Contractor may make use of the quotation for any other purpose; for instance, the Employer may not use the quotation and details to try and obtain a quote from other contractors.

Time Periods

7. The time periods specified in clauses 5.3.1 and 5.3.2 or Schedule 2 may be altered with the agreement of the Employer and Contractor. The parties may agree to extend the specified periods or shorten them. The Employer must confirm any such agreement in writing to the Contractor.

Schedule 3: Insurance Options

This Schedule relates to the Works insurance requirements that are identified in clause 6.7. A further explanation of the Insurance Options and policy issues is contained within the following footnotes:

Footnote [63]   Insurance Options A and B are for use where the work is to be the erection of a new building. Where Option A is chosen, the Contractor is responsible for taking out a Joint Names Policy for all-risks insurance, and where Option B is chosen, the Employer takes out for the Joint Names Policy for all-risks insurance. Where the work comprises alterations to an existing structure and/or an extension to an existing structure, Option C must be used. Under Option C the Employer is responsible for taking out a Joint Names Policy for all-risks insurance for the Works, and a Joint Names Specified Perils insurance for the existing building and for the contents owned by the Employer, or for which he is responsible. It is advised that some Employers may have difficulty in obtaining Joint Names cover for the Option C insurances and in particular the Specified Perils insurance cover. In such a situation the contract will have to be amended, e.g. Option C should not be used and an alternative approach will have to be adopted. The JCT Guide[4] provides some initial advice on this matter, see paragraphs 70, 82 and 83.

Footnote [64] It is important to be aware that there is not a standard All Risks policy that is issued by all insurers. The wording of an All Risks policy and its cover may vary from insurer to insurer. Clause 6.8 describes the level of cover required under this contract, but the wording of some policies may not quite match these requirements and, in some instances, certain types of cover may not be available, e.g. Terrorism Cover. Any discrepancies between the requirements of clause 6.8 and the proposed All Risks Insurance policy should be discussed and agreed between the Employer and Contractor prior to entering into a binding contract. Also see similar advice in Footnote [55].

Footnote [65] This footnote advises the reader to refer to the JCT Guide to obtain further information relating to the reinstatement value of the Works, irrecoverable VAT and other possible costs. The advice provided in the Guide is as follows:

The reinstatement value should include the cost of any professional fees and the cost of removing any debris. Some insurers automatically allow for the cost of removing debris in their All Risks policy but others do not, therefore it is important to check that an adequate level of cover is being provided by the chosen policy.

When the Contractor carries out the reinstatement works he will have to charge VAT on the supplies provided to the Employer, i.e. on the cost of labour, goods and materials supplied. If the Employer is unable to reclaim this VAT, because he is not VAT registered or the supplies are not recoverable under VAT regulations, it is important to include this cost when calculating the reinstatement value of the works.

The All Risks policy provides cover for loss or damage to the Works, etc. but the Employer needs to consider other losses that he may incur as a result of damage to the Works. For example, the Employer may incur commercial losses through the Works being delayed and may also incur higher building costs because of a delay to the Works. Consideration should be given whether to obtain additional insurance to cover these risks.

Insurance Option A

New Buildings – All Risks Insurance of the Works by the Contractor
Contractor to Take Out and Maintain a Joint Names Policy

A.1 The Contractor is required to take out and maintain the All Risks Insurance for the full reinstatement value of the Works or Sections. The policy must be provided by an insurer approved by the Employer, and the cover provided should comply with the requirements of clause 6.8. The policy should also include an allowance to cover professional fees. The wording within this paragraph implies that an allowance for professional fees will be required only where a percentage to cover professional fees has been included in the Contract Particulars. However, if no percentage is actually inserted in the Contract Particulars to cover professional fees, it is stated that the figure of 15% will automatically apply (have the JCT perhaps inadvertently retained some of the wording from the previous edition of the contract?). The Contractor is to maintain the insurance until practical completion, or until the date of termination of the Contractor's employment, whichever is the earlier. The Contractor's obligation to insure ceases even if the validity of the termination of the Contractor's employment is challenged. If the Employer takes over part of the Works through partial possession (see clause 2.36), the Contractor is no longer responsible for insuring that part.

Where a Section Completion Certificate is issued for a relevant Section, the Contractor is no longer obliged to maintain the Joint Names All Risks policy for that Section.

Insurance Documents – Failure by Contractor to Insure

A.2 With reference to the Joint Names Policy referred to in paragraph A.1, the Contractor must send the policy, plus any endorsements and premium receipts, to the A/CA, who will deposit them with the Employer. If the Contractor fails to provide or maintain the necessary insurance cover, the Employer *may* take out a Joint Names Policy for any risks that have not been insured, and recover the costs of the premium(s) from the Contractor. The Employer has the same right of action if the Contractor fails to maintain his annual All Risks policy detailed in paragraph A.3. The Employer may recover the cost of any premiums from monies due to the Contractor, by issuing a withholding notice against an interim payment, or as a debt, e.g. invoice the Contractor for the monies.

Use of Contractor's Annual Policy – as Alternative

Many contractors will hold an annual All Risks policy as part of their normal business operations. This clause details how such a policy may be used to satisfy the requirements of this contract.

A.3 If a Contractor holds and maintains an insurance policy for the Works or Sections which meets the following criteria:

A.3.1 it provides All Risks Insurance which meets the requirements of paragraph A.1 with regard to the level of cover, the full reinstatement value and allowance for professional fees; and

A.3.2   the insurance is a Joint Names Policy, i.e. both the Contractor and Employer are named on the policy.

Where a Contractor provides an annual All Risks Insurance which meets the above criteria, then he will be deemed to have satisfied his obligations to provide insurance under paragraph A.1. The Employer is entitled to inspect the policy and premium receipts at any reasonable time, or the Employer may request the Contractor to send the policy and premium receipts to the A/CA for inspection. As long as the Contractor provides the reasonably requested documentary evidence, the policy and premium receipts may not be retained by the Employer and must be returned to the Contractor. At the pre-contract stage the Contractor will be required to provide the Employer with the annual renewal date of the policy so that the date may be stated in the Contract Particulars.

Loss or Damage, Insurance Claims and Contractor's Obligations

A.4.1   On first discovering that the Works or Site Materials have been lost or damaged as the result of an 'All Risks' event, the Contractor is to immediately give a written notice to both the A/CA and Employer informing them of the nature, extent and location of the damage.

A.4.2   In this paragraph reference is made to clause 6.10.4.2, which relates to the Employer's obligation to pay for damage caused through terrorism, and paragraph A.4.4, which relates to the payment of insurance monies for the damaged works. Despite these payments, the Contractor is still entitled to receive payment (i.e. through Interim Certificates) for the original Works or Site Materials that have been lost or damaged. For example, when an Interim Certificate is issued it must include any work properly carried out by the Contractor, even though it may now have been totally destroyed by fire. The Contractor will also be paid for restoring the damaged work under a certificate of the A/CA. In effect, the Contractor is paid twice for the same work; the original value of the work executed and then the restoration of the work.

A.4.3   If the insurers wish to inspect the Works in response to a claim under the All Risks Insurance the Contractor must wait for this to be completed before he starts to restore the damaged works, etc. After the inspection the Contractor is to restore damaged work, replace damaged items, remove any associated debris in a timely fashion and proceed with the Works.

A.4.4   The Contractor (on his own behalf and also on behalf of any sub-contractors recognised under the All Risks policy) is required to authorise the insurer to pay the insurance money directly to the Employer. The Employer must then pay all the insurance monies to the Contractor by instalments, i.e. the A/CA will issue 'reinstatement' certificates at the same time as he issues Interim Certificates. The Employer is entitled to withhold, from the insurance monies, any expense he has properly incurred in professional fees and in accordance with paragraph A.4.5.

A.4.5   When paying the insurance money to the Contractor, the Employer may withhold any expense properly incurred in respect of professional fees.

This deduction may not exceed the amount calculated by applying the percentage allowed for professional fees (stated in the Contract Particulars) to the net amount paid to the Contractor. For example, a project has suffered from fire damage and £105,000 of insurance monies is due to be released to the Contractor but at the same time the Employer has incurred professional fees of £6,000. The Contract Particulars have been filled in to show a 5% allowance for professional fees. In this situation the Employer will have to release £100,000 to the Contractor and retain 5% of that figure (i.e. £5,000) to cover the professional fees. The Employer may be able to recover the £1,000 shortfall if the insurance company is to make further payments for the fire damage.

A.4.6 The Contractor's reimbursement for the repair work, etc. is limited to the insurance monies received. If there is a shortfall in the insurance monies because of underinsurance, etc. the Contractor will have to make good that shortfall.

Terrorism Cover – Premium Rate Change

A.5.1 If the cost of providing Terrorism Cover (as required in the Joint Names All Risks Insurance) is varied at renewal date, the Contract Sum will be adjusted accordingly. Footnote [55] advises that, where Terrorism Cover is provided, it is done so by means of an additional premium; therefore if, on the renewal of the All Risks policy, the additional premium for Terrorism Cover has been increased by £2,000, the Contract Sum must be increased by £2,000.

A.5.2 Where the Employer is a Local Authority they may, if they wish, comply with the procedure in paragraph A.5.1 but they are allowed to take a different approach to changes in the cost of Terrorism Cover. Under this paragraph the Employer may instruct the Contractor not to renew the Terrorism Cover, which means that from the renewal date the Works will not be covered for damage from an act of terrorism. If the Works should be subsequently damaged through an act of terrorism, the procedures in clauses 6.10.4.1 and 6.10.4.2 would apply, i.e. the Contractor must restore the Works and the Employer would be liable for the cost.

Insurance Option B

New Buildings – All Risks Insurance of the Works by the Employer
Employer to Take Out and Maintain a Joint Names Policy

This clause is to be used where the Employer takes out the All Risks Insurance. For example, some Employer organisations who are regularly involved in construction work may consider they can obtain better or cheaper insurance cover than the Contractor. The obligations on the Employer very much mirror the Contractor's obligations where paragraph A.1 is operative.

B.1 This clause is nearly identical to paragraph A.1, except that it is the Employer who is responsible for taking out and maintaining a Joint Names Policy for the all risks insurance.

Evidence of Insurance

B.2.1   It is identified here that the next two sub-paragraph sections apply only where the Employer is *not* a Local Authority.

B.2.1.1   The Contractor may reasonably require the Employer to produce documentary evidence and receipts to confirm that the appropriate insurance is in place.

B.2.1.2   If the Employer fails to properly take out the insurance (either entirely or in part) the Contractor *may* take out a Joint Names Policy to cover any risk for which the Employer is in default. The cost of any premiums will be added to the Contract Sum.

B.2.2   Where the Employer is a Local Authority the Contractor may reasonably require the Employer to produce a copy of the cover certificate issued by the Joint Names insurer, certifying that Terrorism Cover is being provided under the policy.

Loss or Damage, Insurance Claims, Contractor's Obligations and Payment by Employer

B.3.1   On first discovering that the Works or Site Materials have been lost or damaged as the result of an 'All Risks' event, the Contractor is to immediately give a written notice to both the A/CA and Employer informing them of the nature, extent and location of the damage.

B.3.2   In this paragraph reference is made to clause 6.10.4.2, which relates to the Employer's obligation to pay for damage caused through terrorism, and paragraph B.3.5 which relates to the payment of insurance monies for the damaged works. Despite these payments, the Contractor is still entitled to receive payment (i.e. through Interim Certificates) for the original works or Site Materials that have been lost or damaged. For example, when an Interim Certificate is issued it must include any work properly carried out by the Contractor, even though it may now have been totally destroyed by fire. The Contractor will also be paid for restoring the damaged work. In effect, the Contractor is paid twice for the same work; the original value of the work executed and then the restoration of the work.

B.3.3   If the insurers wish to inspect the Works in response to a claim under the All Risks Insurance, the Contractor must wait for this to be completed before he starts to restore the damaged works, etc. After the inspection the Contractor is to restore damaged work, replace damaged items, remove any associated debris in a timely fashion and proceed with the Works.

B.3.4   The Contractor (on his own behalf and also on behalf of any sub-contractors recognised under the All Risks policy) is required to authorise the insurer to pay the insurance money directly to the Employer.

B.3.5   The restoration work is to be paid through Interim Certificates as though the work were a variation, the implication being that the Employer will have to pay for all of the restoration work even if there is a shortfall in the insurance monies received.

Insurance Option C

Insurance Option C is a complex insurance provision because it deals with a situation where there is a mix of existing property belonging to the Employer, and new works, or alteration works, which are in the control of the Contractor. Without careful management there is potential for insurance disputes to arise as to who is responsible for insuring the various elements of the work and the type of cover being provided. In the past there have been disputes concerning the level of cover provided by this option because of the complexity and possible ambiguity of the wording used in the contract.

(See Ossory Road (Skelmersdale) Ltd v Balfour Beatty Building Ltd, 1993)

Insurance by the Employer of Existing Structures and Works in or Extensions to Them
Existing Structures and Contents – Joint Names Policy for Specified Perils

C.1   The Employer is required to take out a Joint Names (Specified Perils) Policy to cover the existing property and contents that he owns, or for which he is responsible. If, and when, the contract Works are taken over by the Employer through partial possession, the value of that work is to be added to the policy. The policy must be for the full reinstatement cost of any damage caused by one or more Specified Perils (see Footnote [65]). The Employer must maintain the policy up to the date of practical completion, or date of the termination of the Contractor's employment, whichever occurs first. The Contractor is required to authorise the insurer to pay, direct to the Employer, any insurance monies arising from a claim.

Note that this insurance only provides cover for Specified Perils (see definition in clause 6.8), and sub-contractors have no benefit under this policy. Sub-contractors need to be aware that they must have an insurance policy in place to cover any damage that they may cause to the Employer's existing building or the Employer's property.

(See British Telecommunications plc v James Thomson & Sons Ltd, 1998)

The Works – Joint Names Policy for All Risks

C.2   The Employer is required to take out and maintain a Joint Names All Risks Insurance that provides the level of cover set out in clause 6.8 (see Footnote [63]). The insurance is to provide for the full reinstatement value of the Works, or Sections, plus the percentage to cover professional fees stated in the Contract Particulars (see Footnote [65]). The Employer is to maintain the insurance until the date of issue of the Practical Completion Certificate, or until the date of termination of the Contractor's employment (regardless of whether the validity of the termination is challenged), whichever is the earlier. If the Employer takes over part of the Works or Section through partial possession (see clause 2.36), his responsibility to insure that part of the work, under paragraph C.2, ceases. The Employer must now ensure that the value of the Works taken over through partial possession is included in his Specified Perils insurance under paragraph C.1.

There is also a confirmation that after the date of issue of a Section Completion Certificate the Employer's insurance obligations under paragraph C.2 no longer apply to that Section.

Evidence of Insurance

C.3.1   It is identified here that the next two sub-paragraph sections apply only where the Employer is *not* a Local Authority.

    C.3.1.1   The Contractor may reasonably require the Employer to produce documentary evidence and receipts to confirm that the insurances required by paragraphs C.1 and C.2 are in place.

    C.3.1.2   If the Employer fails to properly take out the insurance (either entirely or in part) the Contractor *may* take out a Joint Names Policy to cover any risk for which the Employer is in default. With reference to the insurance for the existing property (paragraph C.1), etc. the Contractor has a right of entry to allow a survey and inventory to be carried out prior to taking out the insurance.

    C.3.1.3   If the Contractor takes out insurance(s) in accordance with paragraph C.3.1.2 then the cost of any premiums will be added to the Contract Sum.

C.3.2   Where the Employer is a Local Authority, the Contractor may reasonably require the Employer to produce a copy of the cover certificates issued by the Joint Names insurer(s) providing cover for paragraphs C.1 and C.2, certifying that Terrorism Cover is being provided under the policies.

Loss or Damage to Works – Insurance Claims and Contractor's Obligations

This section deals with the procedures to be followed if the Works are damaged by an All Risks event and relates therefore to the insurance taken out in compliance with paragraph C.2. This section provides no procedures or advice in a situation where the Employer's existing building or property is damaged by a Specified Perils event.

C.4.1   On first discovering that the Works or Site Materials have been lost or damaged as the result of an 'All Risks' event, the Contractor is to immediately give a written notice to both the A/CA and Employer informing them of the nature, extent and location of the damage.

C.4.2   The Contractor will receive payment for the restoration of damage in accordance with paragraph C.4.5.2 and clause 6.10.4.2. The Contractor is also entitled to receive payment (i.e. through Interim Certificates) for the original works or Site Materials that have been lost or damaged. For example, when an Interim Certificate is issued it must include any work properly carried out by the Contractor, even though it may now have been totally destroyed by fire.

C.4.3   The Contractor (on his own behalf and on behalf of all sub-contractors recognised under the All Risks policy) is required to authorise the insurer to pay the insurance money directly to the Employer.

C.4.4 Within 28 days of any damage to the contract Works or Site Materials caused by an insurable risk it is possible for the Contractor's employment to be terminated, if this action is thought to be just and equitable. It should be noted that this clause does not apply to damage caused to the existing property covered by insurance under paragraph C.1. The termination notice may be given by either party, by actual delivery, special delivery or recorded delivery. Where a termination notice is given the following may apply:

C.4.4.1 if the recipient of the notice considers the termination is not '*just and equitable*' they have 7 days from receipt of the notice to refer the dispute to adjudication; and,

C.4.4.2 where a notice of termination has been given, or where the termination has been disputed and subsequently upheld, clauses 8.12.2 to 8.12.5 (except clause 8.12.3.5) shall apply. Clauses 8.12.2 to 8.12.5 identify the procedures to be followed and the payments to be made following the termination of the Contractor's employment. The exclusion of clause 8.12.3.5 means that the Contractor is not entitled to claim any direct loss and/or damage caused by the termination.

C.5 If no notice of termination is served under paragraph C.4.4, or where the termination has been disputed and the objection upheld, then the following will apply:

C.5.1 after any inspection by the insurer for the All Risks policy has been completed (this does not include the existing property, etc.) the Contractor should restore or replace the damaged work, dispose of any debris in a timely manner and proceed with the Works; and

C.5.2 the Contractor will recover the cost of the above restoration work, etc. by payment through Interim Certificates; the work is to be valued as though it were a variation (see clause 5.2.1.2).

Schedule 4: Code of Practice

Schedule 4 and the following Code of Practice should be read in conjunction with clause 3.18.4. This clause relates to a situation where it is discovered that the Contractor has provided work and/or materials not in accordance with this contract, and the A/CA has concerns that the same 'defective' work may have been executed elsewhere within the Works. Through clause 8.4.4 the A/CA may instruct the Contractor to open up further areas of the Works to allow for inspections and/or tests to discover whether further similar defective work exists. These tests, etc. are at the Contractor's expense and, as a result, the Contractor and A/CA may have widely differing views as to the type and extent of inspection or testing that needs to be carried out. Any instruction given by the A/CA under clause 3.18.4 for the Works to be opened up, etc. must be fair and reasonable. The JCT has produced this Code of Practice to help the A/CA to operate the provisions of clause 3.18.4 in a fair and reasonable manner.

The A/CA should try and reach an agreement with the Contractor upon the amount and method of opening up, inspecting and testing. The A/CA does not have to get the

Contractor's agreement before he issues his instruction, but he must be able to demonstrate that he tried to reach an agreement with the Contractor. When issuing his instruction the A/CA must take into consideration the following criteria:

1. the need to be able to demonstrate, at no cost to the Employer, that the Contractor's non-compliant work was a one-off event, or to demonstrate the extent of similar non-compliant work that exists in Works already constructed or yet to be constructed;
2. if the non-compliant work relates to a primary structural element, it is necessary to identify whether the failing is a result of poor workmanship and/or materials and whether such failure would require a rigorous testing of similar elements. If the non-compliant work relates to a less significant element, it is necessary to identify whether the failure is of a type that would be statistically expected (i.e. a type of work where it is expected that there will be a standard percentage failure) and may be simply repaired. Finally, it must be identified whether the non-compliant work is indicative of an inherent weakness that can be discovered only through selective testing. The extent of such testing would depend upon the importance placed upon the defective work.

    To assist in the operation of the above, the following points from the Code of Practice should be considered:

3. the significance of the non-compliant work in relation to the type of work in which it occurred: for example, the discovery that the backs of skirting boards had not been primed prior to fixing may be an example of non-compliant work but is not a particularly serious omission. Whereas, if it were discovered that external weatherboarding had not been pressure impregnated with preservative prior to fixing this could adversely affect its ability to withstand rot, and is a more serious omission;
4. if there were to be further non-compliant work how would this affect the safety of the building with reference to the occupiers, adjoining buildings and members of the public? Consideration must also be given to compliance with any Statutory Requirements, e.g. CDM Regulations, gas and electricity regulations;
5. consideration should be given to the level and standard of supervision and control of the Works by the Contractor. For example, what supervisory staff are employed upon the site, their background, experience and qualifications, what is the general standard of work achieved by the Contractor to date?
6. any relevant records of the Contractor or his sub-contractors should be reviewed. For example, where a Contractor maintains full and accurate site records it may be easier to pinpoint specific areas of work that may require further testing, or identify why the non-compliant work was executed in the first place;
7. consideration should be given to codes of practice or advice issued by a responsible body, which are relevant to the non-compliant work. For example, the British Woodworking Federation provide advice and guidance on the installation and glazing of timber windows;
8. has the Contractor failed to carry out any tests specified in the contract or in an A/CA instruction? For example, the SMM7 has testing as a measurable item for a

number of trades such as: concrete piling, diaphragm walling, drainage, transport systems, mechanical and electrical services;

9. why was the work carried out in such a way that it failed to meet the contract standards? For example, was it a result of poorly detailed (or lack of) working drawings or specification, poor site supervision, a misunderstanding, careless work, etc.?

10. consideration should be given to any technical advice the Contractor has been given in relation to the non-compliant work. For example, a materials manufacturer may be able to offer a testing service or advice on how to identify or test for non-compliant materials;

11. what are the currently recognised testing procedures in relation to the non-compliant work?

12. what is the practicability of adopting a progressive testing procedure to determine the likelihood of further non-compliant work? For example, specific areas may be chosen for substantial testing and, if these prove satisfactory, then other areas may be chosen for less intensive testing; if these in turn prove satisfactory one or two further spot tests may be carried out;

13. if there are a number of different testing procedures available, then consideration should be given to the time and cost implications of each testing procedure;

14. consideration should be given to any proposals put forward by the Contractor;

15. the final piece of advice is that the A/CA should consider any other relevant matters. Consequently, the A/CA should not limit himself solely to the matters identified above but must be aware of any issues that may have a bearing on the non-compliant work.

Example:

It has been noticed that a laminated timber beam is showing signs of excessive deflection. This beam is a primary structural element and it is therefore important to determine how rigorous and extensive the testing should be for similar elements. From an inspection of the element there is no obvious failing in the workmanship, but there is a problem with the material as the beam appears to be delaminating (re. paragraph 2). If there are similar defective beams it could have serious consequences for health and safety (re. paragraphs 3 and 4). The manufacturer of the laminated beams has confirmed that the beam is failing because of a manufacturing problem in the resin bonding process (re. paragraphs 9 and 10). The manufacturer has identified the production run potentially affected by this fault and confirmed that 5 potentially defective beams had been delivered to site. From site records the Contractor is able to pinpoint the potentially defective beams (re. paragraphs 6 and 9). The manufacturer is able to provide a non-destructive method of testing the beams to identify whether or not they are defective (re. paragraph 10). The Contractor has put forward a proposal to test the 5 beams that have been clearly identified as being potentially defective (re. paragraph 14). The A/CA accepts this proposal but also requires the spot testing of a small number of other beams (re. paragraph 12). If this procedure were to be followed, it should be apparent that the instruction of the A/CA is reasonable in light of the circumstances.

Schedule 5: Third Party Rights

This Schedule relates to the giving of specific rights under this contract to third parties. These rights are given under the Contracts (Rights of Third Parties) Act and this Schedule should be read in conjunction with clauses 7A and 7B of the contract conditions and Part 2 of the Contract Particulars.

Preliminary – Definitions

Clarification is provided here to explain that, where a reference is made to 'Consultants' and 'Sub-Contractors' in Schedule 5, they are the persons identified in Part 2 of the Contract Particulars under the heading 'P & T Rights Particulars' and 'Funder Rights Particulars'.

Part 1: Third Party Rights for Purchasers and Tenants

The following paragraphs set out the administrative procedures and contractual detail relating to the giving of third party rights to Purchasers and Tenants.

1.1 The Contractor warrants that he has carried out the Works, and/or Sections, in accordance with the contract. This warranty from the Contractor becomes effective from practical completion of the Works, or practical completion of a relevant Section. Where the Contractor fails to meet his obligations under the warranty, he shall be liable (subject to the terms in paragraphs 1.2, 1.3 and 1.4) for the following:

    1.1.1 the Contractor is responsible for the reasonable cost of repair, renewal, etc. of any part(s) of the Works. The Contractor's financial liability is limited to the costs incurred by the Purchaser or Tenant and/or the Purchaser's or Tenant's liability to make a financial contribution for such costs; and,

    1.1.2 in addition to the costs above the Contractor may also be responsible for paying any other losses (up to a maximum limit stated in the Contract Particulars) incurred by the Purchaser or Tenant. The Contractor is liable to pay these additional costs only if the Contract Particulars state that paragraph 1.1.2 is to apply and a maximum financial liability has been stated.

    Note – if no financial limit is stated in the Contract Particulars it is not possible to enforce paragraph 1.1.2 (see Contract Particulars Part 2).

1.2 This is a reiteration of the Contractor's liability under paragraph 1. If the Contract Particulars state that paragraph 1.1.2 is not to apply, or it is deemed not to apply (e.g. no information provided on maximum liability), then the Contractor is liable only for the costs identified in paragraph 1.1.1.

1.3 This section identifies the fact that the Contractor is responsible to the Purchaser and Tenant only for losses that are directly attributable to the Contractor. For example, a Purchaser may have incurred losses of £100,000 because of a combination of poor workmanship by the Contractor and poor design detailing by the A/CA. The Contractor's poor workmanship is responsible for 40% of the losses. The Contractor's financial liability is therefore £40,000; clauses such as this are commonly referred to as 'net contribution clauses'. The net contribution payment in this paragraph is based upon the assumptions detailed below:

1.3.1 either the consultant has provided a contractual undertaking (e.g. a collateral warranty) with a Purchaser or Tenant in respect of the standard of work that he has performed for the Employer in accordance with the terms of his consultancy agreement, and there are no limitations on the consultant's liability within the consultancy agreement for the Works; or alternatively the consultant has provided the same rights to a Purchaser or Tenant through the Contracts (Rights of Third Parties) Act. This means the consultant, by one of two legal means, owes the same liability to the Purchaser and Tenant as he owes to the Employer under the consultancy agreement for the Works;

1.3.2 this is a similar clause to the above to the extent that the sub-contractor has provided legal rights to a Purchaser or Tenant in respect of the design work carried out by the sub-contractor for the sub-contract works. There is an important caveat to say that this paragraph is relevant only to design work carried out by the sub-contractor and for which the Contractor has no liability to the Employer under this contract. In normal circumstances, a Contractor would be liable to the Employer for a sub-contractor's design, see clauses 3.7.2, 2.2 and 2.19;

1.3.3 the consultant and sub-contractor have paid a proportion of the losses incurred by the Purchaser or Tenant. This proportional payment is to be just and equitable and is based upon the extent of the consultant's and/or sub-contractor's responsibility for the Purchaser or Tenant loss;

It is important to understand that the events in paragraphs 1.3.1 to 1.3.3 do not have to actually take place, but it is assumed they have taken place to enable the Contractor's proportion of the losses to be justly and equitably determined.

1.4 Where a Purchaser or Tenant starts an action or proceedings against the Contractor, the Contractor may use any of the terms in the contract in his defence. Whatever rights the Contractor may possess in an action with the Employer, he may exercise the same rights in an action with the Purchaser or Tenant.

1.5 If the Purchaser or Tenant appoints a person to carry out an independent investigation into any relevant matter, this will do nothing to remove or diminish the Contractor's obligations under paragraph 1, therefore if a Purchaser or tenant has a structural survey carried out to determine the cause of a building defect, this will not alter the Contractor's obligations.

2. The Contractor warrants that (unless authorised) he has not used, and will not use, materials that do not comply with the guidelines (current at the date of contract) contained within the 'Good Practice in Selection of Construction Materials'

produced by Ove Arup. The Contractor, without breaching his warranty, may use materials that do not comply with the guidelines where they are required by the contract, or their use has been authorised by the Employer or A/CA. This authorisation should be given in writing but, if given orally, this must be confirmed by the Contractor in writing. Where the Contractor is in breach of the warranty provided by paragraph 2, then the provisions of paragraph 1 will apply.

3. This paragraph explains that there is no procedure or authority for the Purchaser or Tenant to issue any direction or instruction to the Contractor with reference to this contract.

4. This paragraph relates to where the Works contains a Contractor's Design Portion and a Purchaser's or Tenant's property contains the CDP works or part of it. As long as the Contractor has been paid all the monies to which he is entitled under this contract, a Purchaser or Tenant has the same rights and licence with regard to the use of the Contractor's Design Documents as does the Employer under clause 2.41 (i.e. Copyright and use). However, the Purchaser or Tenants rights and licence are subject to the same limitations and exclusions placed upon the Employer through clause 2.41.

5. Where the Works includes a Contractor's Design Portion, the Contractor warrants that he has, and shall maintain, a professional indemnity insurance in accordance with clause 6.11 and in compliance with the details set out in the Contract Particulars. The Purchaser or Tenant has the right to ask the Contractor to produce documentary evidence to show that the professional indemnity insurance is being maintained. The request to inspect the documentation is to be made on a reasonable basis. If the insurance ceases to be available at commercially reasonable rates, the Contractor must immediately notify the Purchaser or Tenant. Following this notification the Contractor and Purchaser or Tenant should discuss how they may best protect their interests in the absence of the professional indemnity insurance.

6. The P&T Rights contained within Part 1 of this Schedule may be assigned, without the Contractor's consent, through the means of an 'absolute legal assignment'. A Purchaser or Tenant may assign their rights under this Schedule to another party, and this party may subsequently assign their rights to another party; therefore, although a Contractor provides third party rights to the original purchaser or tenant, he must be aware of the mechanism to assign these rights to subsequent Purchasers or Tenants. The P&T Rights relating to a Purchaser or a Tenant may only be assigned twice. Although an assignment may take place without the Contractor's consent, it will not be effective until the Contractor is given a written notice of the assignment.

   A brief explanation of an absolute legal assignment relating to property is provided in the Law of Property Act 1925. The assignment must be complete; it is not allowable to pass on some of the rights or to try and qualify the rights that are being assigned. The assignor must give a written notice of the assignment and having given the assignment, the assignor has no rights to sue the Contractor under this warranty.

7. This paragraph sets out the manner in which notices are to be delivered. Any notice from the Contractor is deemed to have been 'duly given' where it is

delivered to the Purchaser's or Tenant's registered office. The means of delivery may be by hand, special delivery or recorded delivery. Notices given by the Purchaser or Tenant are governed by the same delivery procedures and are to be delivered to the Contractor's registered office. Where a notice is sent by special or recorded delivery, it is deemed to have been received 48 hours from time of posting, unless it can be proved otherwise.

8. This paragraph specifies the period of time for which the Contractor will be liable to Purchasers and Tenants under the P&T Rights. The period of liability runs from the date of practical completion of the Works or alternatively, where the Works are broken down into Sections, the period of liability will run from the date of practical completion of each Section. The date the Contractor's liability is extinguished depends upon the following:

    8.1 if the Contract is executed under hand, the Contractor's liability will cease 6 years from the date of practical completion of the Works or the relevant Section;

    8.2 if the Contract is executed as a deed, the Contractor's liability will cease 12 years from the date of practical completion of the Works or the relevant Section.

9. This paragraph confirms that the Contractor cannot be liable to a Purchaser or Tenant if he fails to complete the Works by the Completion Date.

10. Any dispute between the Contractor and Purchaser or Tenant concerning the P&T Rights will be controlled by the law of England, and English courts will have jurisdiction over the dispute or difference.

Part 2: Third Party Rights for a Funder
The following paragraphs set out the administrative procedures and contractual details relating to the giving of third party rights to a Funder, i.e. an organisation that has lent money to the Employer to enable the project to proceed.

1. In this paragraph the Contractor provides the Funder with a warranty that he has complied and will comply with this contract. If the Contractor is found to be in breach of this warranty the following sub-sections become relevant:

    1.1 The Contractor's liability for the Funder's losses is limited. The Contractor is liable to the Funder only for losses that can be fairly attributed to the Contractor's breach of the warranty. To give an example, a Funder is facing costs of £500,000 as a result of a number of failings within a project; from an examination of the facts it has been assessed that the Contractor and A/CA are equally responsible for the losses, therefore the Contractor's financial contribution is limited to 50%, i.e. £250,000. A just and equitable assessment of the Contractor's proportional liability must be based on the following assumptions:

        1.1.1 the consultant has provided a contractual undertaking (e.g. a collateral warranty) with the Funder to the effect that he has performed or will perform his services, in connection with the Works, in accordance with the terms of his consultancy agreement with the Employer; there are no limitations on the consultant's liability within the consultancy agreement for the Works. Alternatively, the consultant has provided the same rights to the Funder through the Contracts (Rights of Third Parties) Act;

1.1.2 this is a similar paragraph to the above to the effect that the sub-contractor has provided legal rights (a collateral warranty or third party rights) to the Funder in respect of the design work carried out by the sub-contractor for the sub-contract works. There is an important caveat to say that this paragraph is relevant only to design work carried out by the sub-contractor for which the Contractor has no liability to the Employer under this contract. In normal circumstances, a Contractor would be liable to the Employer for a sub-contractor's design, see clauses 3.7.2, 2.2 and 2.19;

1.1.3 the consultant and sub-contractor have paid a proportion of the losses incurred by the Funder. This proportional payment is to be just and equitable and is based upon the extent of the consultant's and/or sub-contractor's responsibility for the Funder's loss.

It is important to understand that the events in paragraphs 1.1.1 to 1.1.3 do not have to actually take place, but it is assumed they have taken place to enable the Contractor's proportion of the losses to be justly and equitably determined.

2. The Contractor warrants that (unless authorised) he has not used and will not use materials that do not comply with the guidelines (current at the date of contract) contained within the 'Good Practice in Selection of Construction Materials' produced by Ove Arup. The Contractor, without breaching his warranty, may use materials that do not comply with the guidelines where they are required by the contract, or their use has been authorised by the Employer or A/CA. This authorisation should be given in writing but, if given orally it must be confirmed by the Contractor in writing. Where the Contractor is in breach of the warranty provided by paragraph 2, then the provisions of paragraph 1 will apply.

3. The Funder normally has no authority to issue the Contractor with instructions or directions under this contract. However, the Funder is entitled to issue the Contractor with instructions or directions under paragraph 5, i.e. where the Funder terminates the Finance Agreement, or under paragraph 6.4, where the Contractor has the opportunity to terminate his employment but the Funder steps in to replace the Employer.

4. The Funder normally has no responsibility to the Contractor for the monies that the Contractor receives under this contract. However, this situation changes once the Funder gives the Contractor a notice under paragraph 5 or 6.4. After one of these notices has been issued, the Funder does become liable to the Contractor for monies that are due under this contract.

5. Where the Funder has terminated the Finance Agreement with the Employer, the Contractor agrees that he will accept instructions from the Funder (or person

appointed by the Funder) when required to do so in a written notice from the Funder. From this point the Contractor is not to accept any instructions from the Employer. The Contractor's agreement to accept the Funder's instructions is subject to the provisions contained within paragraph 7. Furthermore, the instructions must relate to the Works and be issued in accordance with the contract conditions. To safeguard the Contractor's position, it is stated that the Employer accepts, for the purpose of this contract, that the Funder's notification to the Contractor may be relied upon by the Contractor as conclusive evidence of the termination of the Finance Agreement by the Funder. The Employer also accepts that the Contractor, by accepting instructions from the Funder to the exclusion of the Employer, shall not be in breach of his obligations to the Employer under this contract.

6.   This paragraph relates to the situation where a Contractor, in accordance with the contract conditions, has the right to terminate his employment. The following sections set out the procedures with which the Contractor must comply before he terminates his employment:

6.1   The Contractor is not to exercise any rights he may have under this contract to terminate his employment (e.g. see clause 8.9) without first providing the following information:

6.1.1   before the Contractor is able to give a notice terminating his employment, he must provide the Funder with copies of the written notices that are required under this contract to be sent to the A/CA or Employer (see clauses 8.9.1, 8.9.2 and 8.11); and

6.1.2   the Contractor must inform the Funder, through a written notice, when he has an immediate contractual right to notify the Employer that his employment is terminated (see clauses 8.9.3, 8.9.4, 8.10.1 and 8.1).

6.2   This paragraph deals with the legal principle of repudiation. For example, if the Employer fails to give the Contractor access to the site on the agreed date, this could be viewed as a repudiation on behalf of the Employer. In response to the Employer's repudiation, the Contractor could treat the contract as though it had been terminated.

If the Contractor does consider that the Employer has repudiated this contract, he must first give the Funder written notice of his intention to inform the Employer of the repudiation.

6.3   This section sets out a timetable for certain notifications. It is a requirement that the Contractor will not do the following:

6.3.1   issue the Employer with any paragraph 6.1.2 notification; or

6.3.2   inform the Employer that he is treating the contract as being repudiated by the Employer as explained in paragraph 6.2 until 7 days after the date the Funder received the Contractor's notification under either paragraph 6.1.2 or 6.2. A period of time other than 7 days may apply if this is stated in Part 2 of the Contract Particulars.

6.4   Before the expiry of the time period set out in paragraph 6.3.2, the Funder may give the Contractor a written notice requiring the Contractor to agree to accept instructions from the Funder (or person appointed by the Funder). Having received such a notice, the Contractor is not to accept any instructions from the Employer. The Contractor's agreement to accept the Funder's instructions is subject to the provisions contained within paragraph 7, also the instructions must relate to the Works and be issued in accordance with the contract conditions. In order to safeguard the Contractor's position it is stated that the Employer acknowledges that the Contractor is 'protected' by the Funder's notice given under paragraph 6.4 and, as a result, the Contractor's acceptance of the Funder's instructions to the exclusion of the Employer may not be viewed as a breach of the Contractor's obligations to the Employer for the purpose of this contract. Finally, it is confirmed that, subject to the provisions in paragraph 7, nothing contained within paragraph 6.4 will relieve the Contractor of his liability to the Employer where the Contractor has been in breach of contract.

7. This paragraph sets out the responsibilities of the Funder where he steps in and takes over the role of the Employer. Where the Funder has given a notice under paragraph 5 or 6.4 it is a requirement that the Funder (or appointee) becomes liable to the Contractor for monies due or payable under this contract. This would also include monies that are outstanding at the date of the notice, i.e. monies that the Employer was obliged to pay but had failed to pay by the due date. The Funder (or appointee) becomes responsible for fulfilling the Employer's role and obligations under this contract. The contract will continue in full force as though the Contractor's right to terminate his employment, or his right to accept the Employer's repudiation of this contract, never existed. Whatever liability the Contractor owed the Employer under this contract is transferred to the Funder (or appointee). Where, in a paragraph 5 or 6.4 notice, the Funder informs the Contractor to accept instructions from an appointee, the Funder remains a guarantor for payments due to the Contractor from the appointee.

8. This paragraph relates to where the Works contains a Contractor's Design Portion. As long as the Contractor has been paid all the monies to which he is entitled under this contract the Funder has the same rights and licence, with regard to the use of the Contractor's Design Documents, as does the Employer under clause 2.41 (i.e. Copyright and use). However, the Funder's rights and licence are subject to the same limitations and exclusions placed upon the Employer through clause 2.41.

9. Where the Works includes a Contractor's Design Portion, the Contractor warrants that he has maintained and shall maintain a professional indemnity insurance in accordance with clause 6.11 and in compliance with the details set out in the Contract Particulars. The Funder or appointee has the right to ask the Contractor to produce documentary evidence to show that the professional indemnity insurance is being maintained. The request to inspect the documentation is to be made on a reasonable basis. If the insurance ceases to be available at commercially reasonable rates, the Contractor must immediately notify the Funder. Following this notification, the Contractor and Funder should discuss how they may best protect their interests in the absence of the professional indemnity insurance.

10. The Funder Rights contained within Part 2 of this Schedule may be assigned, without the Contractor's consent, through the means of an 'absolute legal assignment'. A Funder may assign his rights under this Schedule to any other person who is providing finance in connection with the carrying out of the Works, and this Funder may subsequently assign their rights to another person providing finance for the Works. Consequently, although a Contractor provides third party rights to the original Funder, he must be aware of the mechanism to assign these rights to subsequent Funders. The Funder Rights may only be assigned twice. It is important to note that, although it is not necessary to obtain the Contractor's consent to an assignment, the assignment will not be effective until the Contractor has been given a written notice of the assignment.

A brief explanation of an absolute legal assignment relating to property is provided in the Law of Property Act 1925. The assignment must be complete, it is not possible to pass on some of the rights or to try and qualify the rights that are being assigned. The assignor must give a signed written notice of the assignment. Having given the assignment, the assignor no longer has any rights to sue the Contractor in relation to this warranty.

11. This paragraph sets out the manner in which notices are to be delivered. Any notice from the Contractor is deemed to have been 'duly given' where it is delivered to the Funder's registered office. The means of delivery may be by hand, special delivery or recorded delivery. Notices given by the Funder are governed by the same delivery procedures and are to be delivered to the Contractor's registered office. Where a notice is sent by special or recorded delivery, it is deemed to have been received 48 hours from time of posting, unless it can be proved otherwise.

12. This paragraph specifies the period of time for which the Contractor will be liable to the Funder under the Funder Rights. The period of liability runs from the date of practical completion of the Works or alternatively, where the Works are broken down into Sections, the period of liability will run from the date of practical completion of each Section. The date the Contractor's liability is extinguished depends upon the following:

    12.1 if the Contract is executed under hand, the Contractor's liability will cease 6 years from the date of practical completion of the Works or the relevant Section;

    12.2 if the Contract is executed as a deed, the Contractor's liability will cease 12 years from the date of practical completion of the Works or the relevant Section.

13. Despite the rights given to the Funder under this Schedule, the Contractor is not liable to the Funder for delays under this contract until the Funder issues a paragraph 5 or paragraph 6.4 notice. After the issue of such a notice, the Contractor is liable to the Funder for delays for which the Contractor is responsible. However, the Contractor cannot be required to pay liquidated damages to the Funder for a period of time where the Employer has already deducted liquidated damages. This means a Contractor cannot be required to pay liquidated damages twice for the same delay.

14.1 Any dispute or difference between the Contractor and Funder concerning the Funder Rights will be controlled by the law of England, and English courts will have jurisdiction over the dispute or difference.

14.2 Where a Funder has given the Contractor a paragraph 5 or paragraph 6.4 notice, any subsequent dispute or difference between the parties shall be subject to the provisions of Article 7 (i.e. adjudication) and by Article 8 (arbitration) if it applies. It will be necessary to check the Contract Particulars to discover whether disputes may be referred to arbitration (i.e. Article 8 applies) in which case clauses 9.3 to 9.8 will also apply. One party to the dispute will be the Contractor and the other party may be the Funder, or his appointee or a permitted assignee.

Schedule 6: Forms of Bonds

Note: unless stated otherwise, any reference to a clause in Schedule 6 is referring to a clause within the bond and not the JCT conditions.

These bonds relate to the payment procedures set out in 'Section 4 Payment', and in particular to the provision of an Advance Payment or payment for materials off site or where the Employer uses the option of not deducting Retention. Reference should be made to clauses 4.8, 4.17 and 4.19 respectively.

Part 1: Advance Payment Bond

Part 1 sets out the terms and details for an Advance Payment Bond. The purpose of the bond is to protect the Employer's interests if a Contractor fails to repay an advance payment.

1. This is where the parties to the bond are identified. This will require the name and registered address of the surety to be inserted followed by the name and address of the Employer.
2. Here the Contractor is identified, followed by a description of the contract Works.
3. This clause identifies the value of the advance payment made by the Employer to the Contractor. It will be necessary to insert the value (in figures) of the payment in the space provided. It is stated that the surety will be liable to reimburse the Employer for the advance payment, based on the following terms:

    3.1 when the surety receives a clause 3.2 demand from the Employer, the surety is required to repay the Employer the amount demanded. The surety's liability is limited to the amount of the advance payment identified in clause 3. The Employer is not entitled to demand a figure higher than the advance payment;

    3.2 whenever the Employer issues the surety with a demand under this bond, it must be in the form of a 'completed notice of demand'. This means that the Employer must complete the proforma Schedule provided with the bond. The completion of this Schedule will be conclusive evidence of the Employer's entitlements under this bond. To safeguard the surety's interests, the two Employer signatures on the Schedule must be confirmed as genuine by the Employer's bankers;

3.3   within 5 business days of receiving the demand, the surety must pay the Employer the amount demanded. A business day is defined as being a day on which commercial banks are open for business in London, apart from Saturdays and Sundays, which are specifically excluded.

4.   Payments are to be made under this bond when they are demanded by the Employer. Payment is to be made even where there is a dispute between the Employer and Contractor or where the Employer or Contractor might have valid claims against each other. The implication of this clause is that an advance payment may not be reduced or negated in recognition of other claims or disputes – it must be paid on demand regardless of other events and disputes. A payment made by the surety under this bond is deemed a valid payment for the purposes of this bond; as a result, the surety's liability under this bond is reduced by the value of the payment made.

5.   This clause identifies certain actions that the Employer may take which will have no effect on the surety's obligations and liability under this bond. The Employer is not required to give the surety notice of the actions and does not need the surety's consent for the actions. The Employer actions that are relevant to this clause are identified in the following:

5.1   'a waiver by the Employer of any terms, provisions, conditions, obligations and agreements of the Contractor'; another permitted action is where the Employer fails to make a demand upon, or fails to take action against, the Contractor;

5.2   any modification or change to the Contract; e.g. the Employer and Contractor may agree to alter clauses in the contract conditions, or alter details in the Contract Particulars;

5.3   granting the Contractor an extension of time that does not affect the timetable set out in clause 7.3.

6.   The surety's liability to the Employer commences on the date the advance payment is made to the Contractor. The maximum aggregate liability of the surety is to be inserted into this clause, and would be the value of the agreed advance payment. When issuing the surety with a demand notice, the Employer may not demand more than the amount of the advance payment (see clause 3.1). Under this clause, if the Employer issues more than one demand notice, then the aggregate of those demands may not exceed the limit stated in this clause. The stated limit will be reduced by the amount of advance payments that are repaid by the Contractor. The Employer is to inform the surety of the value of any advance payment that has been repaid by the Contractor. This information is to be given as a written notice.

Example:

The Employer makes an advance payment to the Contractor of £750,000, to be repaid in three equal instalments in month 2, 4 and 6. The surety's maximum aggregate limit is stated as £750,000. The first repayment is made in month 2 (see SBC/Q clause 4.10.2), which results in the surety's liability being reduced to £500,000. The Employer fails to ensure the second repayment is made; the surety's liability remains at £500,000. In month 5 the Contractor goes into liquidation, but the Employer may still claim from the surety for the missed repayment for month 4. The surety's liability would now stand at £250,000. The Employer would be able to claim this sum if the liquidator is unable to make the final repayment of the Contractor's advance payment.

7. This is an important clause because it sets out a timetable that determines when the surety's liability to the Employer ceases. The surety's liability comes to an end from the date of occurrence of the earlier of the following events:

   7.1 the date that the advance payment is reduced to nil. The Employer is to give the surety a written notice confirming the date that this event occurred;

   7.2 the date that the advance payment or the remaining balance is repaid to the Employer by either the Contractor or surety. The Employer is to give the surety a written notice confirming the date that this event occurred;

   7.3 it is a requirement of the bond that a date be inserted into this section, the implication being that the surety's liability under this bond ceases after this date. The Employer would have to be advised as to an appropriate date to ensure that he retains the protection of this bond whilst the advance payment (or balance) remains outstanding.

   If the Employer is to make a claim under this bond, it must be submitted in writing (i.e. the Notice of Demand) and be received by the surety before the earliest of the three above dates.

8. The Employer has no authority to transfer or assign his rights under this bond without the surety's consent. However, the surety cannot withhold his consent unreasonably.

9. The purpose of this section is to clarify the fact that the only parties who may enforce the terms of the bond are the Employer and surety. The wording of the clause is such that it excludes third parties from using the Contracts (Rights of Third Parties) Act to claim the existence of an enforceable benefit within the bond.

10. If there are any legal issues relating to the bond, they are to be resolved in accordance with the laws of England and Wales.

IN WITNESS

The name of the surety is to be inserted in the bond, and the bond is to be signed, as a deed, by a legally authorised representative of the surety. The signing of the bond is to be witnessed and dated.

The significance of signing a contract as a deed is that a party is potentially liable for any breach of contract for 12 years. In this instance the surety's obligations are limited to the timetable identified in clause 7.

Schedule to Advance Payment Bond

This is a proforma provided within the contract to assist in the operation of clause 3.2 of the bond. If the Employer wishes to make a demand on the surety, then this 'Notice of

Demand' must be properly completed. Under this notice the Employer may not claim a figure that is in excess of the repayment which the Contractor has currently failed to make in breach of this contract.

The notice must be signed by two authorised officials of the Employer, and the signatures must be authenticated by the Employer's bank.

Part 2: Bond in Respect of Payment for Off-Site Materials and/or Goods

This is a bond for use by the Employer where it has been agreed in the Contract Documents that the Contractor will be entitled to be paid for certain off-site materials through Interim Certificates. SBC/Q clauses 4.17.4 and 4.17.5 should be referred to in the context of this bond. Clause 4.17.4 relates to off-site materials that are 'uniquely identified' and as such the use of this bond is optional. The Employer will have stated in the Contract Documents whether or not he requires a bond from the Contractor (see Contract Particulars). Clause 4.17.5 relates to off-site materials that are not uniquely identified. Where clause 4.17.5 is operational, then the Contractor is required to provide a bond; it is not optional. This bond may be used for both uniquely identified listed materials (clause 4.17.4) and listed materials that are not uniquely identified (clause 4.17.5).

1. This is where the parties to the bond are identified. This will require the name and registered address of the surety to be inserted, followed by the name and address of the Employer.
2. Here the Contractor is identified, followed by a description of the contract Works.
3. This clause identifies the main provisions within the contract, relating to the payment of off-site materials. The details are as follows:

    3.1 the Employer has agreed to pay (through Interim Certificates) for off-site materials, etc. which have been identified in a list, attached to the contract. For the purposes of this bond the materials, etc. are referred to as 'Listed Items';

    3.2 the Contractor has agreed to insure the Listed Items against loss and damage. The insurance is to be for the full value of the Listed Items and the policy is to protect the interests of both the Employer and Contractor. The policy is to run from the time the property in the Listed Items is transferred to the Contractor until they are delivered to or adjacent to the Works (see SBC/Q clause 4.17.2);

    3.3 this bond relates exclusively to the monies paid to the Contractor by the Employer for Listed Items that are stored off site.

4. If the Employer issues the surety with a demand under this bond, it must be in the form of a 'completed notice of demand'. This means that the Employer must complete the proforma Schedule provided with the bond. The completion of this Schedule will be conclusive evidence of the Employer's entitlements under this bond. To safeguard the surety's interests, the two Employer signatures on the Schedule must be confirmed as genuine by the Employer's bankers.
5. Within 5 business days of receiving the demand, the surety must pay the Employer the amount demanded. A business day is defined as being a day on which commercial banks are open for business in London, apart from Saturdays and Sundays, which are specifically excluded.

6. Payments are to be made under this bond when they are demanded by the Employer. Payment is to be made even where there is a dispute between the Employer and Contractor or where the Employer or Contractor might have valid claims against each other. The implication of this clause is that any claim against the bond, by the Employer, may not be reduced or negated in recognition of other claims or disputes. The claim must be paid on demand regardless of other events and disputes. A payment made by the surety under this bond is deemed a valid payment for the purposes of this bond; as a result, the surety's liability under this bond is reduced by the value of the payment made.

7. This clause identifies certain actions the Employer may take which will have no effect on the surety's obligations and liability under this bond. The Employer is not required to give the surety notice of the actions and does not need the surety's consent for the actions. The Employer actions that are relevant to this clause are as follows:

    7.1 a 'waiver by the Employer of any terms, provisions, conditions, obligations and agreements of the Contractor'; another permitted action is where the Employer fails to make a demand upon, or fails to take action against, the Contractor;

    7.2 any modification or change to the Contract; e.g. the Employer and Contractor may agree to alter clauses in the contract conditions, or alter details in the Contract Particulars;

    7.3 when the Contractor is granted an extension of time that does not affect the timetable set out in clause 9.2.

8. In this clause is to be inserted a sum of money which will be the maximum aggregate liability of the surety; i.e. if the Employer makes a number of claims under this bond, the sum total of the claims may not exceed this stated limit. This maximum limit will be taken from the Contract Particulars. The Employer will need to be advised as to the maximum sums he may need to pay the Contractor for off-site Listed Items.

9. This is an important clause as it sets out a timetable that determines when the surety's liability to the Employer ceases. The surety's liability comes to an end from the date of occurrence of the earlier of the following two events:

    9.1 the date that all Listed Items have been delivered to, or adjacent to, the Works. The Employer is to give the surety a written notice confirming the date when this event has occurred;

    9.2 it is a requirement of the bond that a 'longstop date' be inserted into this section, the implication being that the surety's liability under this bond ceases after this date. This date would have to take into account the Listed Items allowed for in this contract and the Contractor's programme of Works, plus an allowance for any potential delays to the progress of the Works.

    If the Employer is to make a claim under this bond, it must be submitted in writing (i.e. the Notice of Demand) and be received by the surety before the earlier of the two above dates.

10. The Employer has no authority to transfer or assign his rights under this bond without the surety's consent. However, the surety cannot unreasonably withhold his consent.

11. The purpose of this section is to clarify the fact that the only parties who may enforce the terms of the bond are those who are party to the agreement. The wording of this clause is such that it excludes third parties from using the Contracts (Rights of Third Parties) Act to claim the existence of an enforceable benefit within the bond.

12. If there are any legal issues relating to the bond, they are to be resolved in accordance with the laws of England and Wales.

IN WITNESS

The name of the surety is to be inserted in the bond, and the bond is to be signed, as a deed, by a legally authorised representative of the surety. The signing of the bond is to be witnessed and dated.

The significance of signing a contract as a deed is that a party is potentially liable for any breach of contract for 12 years. In this instance, the surety's obligations are limited by the timetable identified in clause 9.

Schedule to Bond

This is a proforma provided within the contract to assist in the operation of clause 4 of the bond. If the Employer wishes to make a demand on the surety, then this 'Notice of Demand' must be properly completed. Under this notice the Employer may claim a figure that equals the payments made for Listed Items (in Interim Certificates) for which the Contractor is in breach by failing to have them delivered on, or adjacent to, the Works. This sum (or aggregate of sums) may not exceed the surety's maximum aggregate liability stated in clause 8. The surety must make payment to the address stated in this schedule.

The notice must be signed by two authorised officials of the Employer, and the signatures must be authenticated by the Employer's bank.

Part 3: Retention Bond

This bond is for use by an Employer who has opted not to deduct Retention from the Contractor in accordance with SBC/Q clause 4.19. The application of this clause and this associated bond is not relevant where the Employer is a local authority. The Contract Particulars will identify whether SBC/Q clause 4.19 applies or not; where it does apply then the Contractor is required to provide a Retention bond.

The initial section of the bond is used to insert the names and addresses of the surety and Employer, and the date of the bond.

1. Here the Contractor is identified by name and address, and it is confirmed that the Employer has agreed not to deduct Retention from payments due to the Contractor where the Contractor has taken out the appropriate Retention bond.

2. In this section is inserted the sum of money for which the surety is liable to the Employer. The value is to be inserted as a figure and in words and will

mirror the amount stated in the Contract Particulars. This figure represents the surety's maximum aggregate liability to the Employer, although this amount is subsequently reduced by 50% from the date of the next Interim Certificate after practical completion. The Employer should inform the surety of the date of issue of the relevant Interim Certificate.

In this clause there is no mention of partial possession or a Section Completion Certificate, both of which result in a release of Retention to the Contractor. However, clauses 4.2 and 4.3 appear to cover these events.

3. As long as the Employer's demand complies with the requirements of clause 4, the surety is obliged to pay the Employer the amount demanded.
4. This identifies the procedures with which the Employer must comply if he is to issue a valid demand under clause 3. The demand shall:

4.1 be in writing and must be addressed to the surety at the address specified within this clause. In this demand the Employer is to refer to this bond, and the signature(s) must be confirmed genuine by the Employer's bankers. Unlike the advance payment bond and off-site materials bond, there is no proforma notice of demand provided for the Retention bond;

4.2 state the amount of Retention that the Employer would have been holding, at the date of the demand, if SBC/Q clause 4.19 was not operational and Retention had been deducted from the Contractor;

4.3 state the amount demanded. The Employer may not demand any more than the figure calculated for clause 4.2 above. The Employer must also identify why he is claiming against the Retention bond. For a claim to be valid, the Employer must identify one or more of the following events as being the reason for the claim:

4.3.1 the actual costs incurred by the Employer because the Contractor failed to comply with an instruction of the A/CA (see SBC/Q clause 3.11). The Employer must also attach a statement from the A/CA confirming the Contractor's failure;

4.3.2 the insurance premiums paid by the Employer to provide insurance that the Contractor was required to take out and maintain, but failed to do (see SBC/Q clause 6.4.1.3 and Schedule 3 Insurance Option A, paragraph A.2).

4.3.3 the liquidated and ascertained damages for which the Contractor is liable. This must be accompanied by the A/CA's Non-Completion Certificate that is relevant to this claim. The certificate may relate to the whole Works or a Section and will confirm the Contractor's failure to complete the Works on time (see SBC/Q clause 2.31).

4.3.4 any expenses or any direct loss and/or damage incurred by the Employer as a result of terminating the Contractor's employment (see SBC/Q clause 8.7.4.1);

4.3.5 this sub-clause gathers up all the other instances where the Employer is entitled, under this contract, to deduct money from the Contractor, i.e. any other claim apart from those already identified in clauses 4.3.1

to 4.3.4. These costs must have been actually incurred by the Employer. The Employer must identify the provision under the contract that entitles him to claim this deduction.

4.4 It is a requirement that the Employer must give the Contractor a written notice informing him why he is being held liable for the amount being claimed under this bond. At the same time the Employer must send a copy of the notice to the surety. The copy must be sent to the surety's address, which will have been inserted in this clause. As part of the clause 4 demand the Employer must certify that he has provided this notice (and a copy to the surety) and that the Contractor has failed to pay the due amount within 14 days of the notice.

If a clause 4 demand complies with all the above conditions, it will be viewed as conclusive evidence (for the purposes of this bond only) that the amount is properly due and payable to the Employer by the Contractor. See 'Note 3' which is appended to the Retention Bond for further comment by the JCT on the wording used for this clause.

5. If the Employer intends to pass the benefits of this contract onto another party through assignment or other legal means, he may only include the benefit of this bond with the surety's consent. The surety's consent must be given in writing and cannot be unreasonably withheld or delayed.

6. As long as the surety has not received a clause 4 demand, his liability comes to an end from the date of occurrence of the earliest of the following three events:

6.1 the date of issue of the Certificate of Making Good (see SBC/Q clause 2.39). It is the Employer's duty to inform the surety of this date;

6.2 the date by which the surety has satisfactorily met the demand(s) of the Employer, up to the maximum aggregate figure allowed under this bond (see clause 2);

6.3 the date inserted in this section of the bond.

7. This clause reinforces the importance of the above timetable of events. Once one of the above events has occurred the bond is terminated and becomes ineffective. If the Employer has a valid claim against the bond he must ensure that the surety receives the clause 4 demand on or before the date of occurrence of the earliest of the above events.

It is important for the A/CA to be aware that his issuing of the Certificate of Making Good may have the effect of terminating this bond. The Employer should also be aware that on occasions it may be advisable to deliver a demand by hand to ensure a surety could not claim the notice was received 'out of time'.

8. The purpose of this section is to clarify the fact that it is only the parties to this bond who may enforce the terms of the bond (i.e. this would initially be the Employer and the surety). The wording of this clause is such that it excludes third parties from using the Contracts (Rights of Third Parties) Act to claim the existence of an enforceable benefit within the bond. However, if the Employer were to properly assign his benefits (see clause 5) the assignee would be able to enforce his benefits.

9.   If there are any legal issues relating to the bond then they are to be resolved in accordance with the laws of England and Wales.

IN WITNESS

The name of the surety is to be inserted in the bond, and the bond is to be signed, as a deed, by a legally authorised representative of the surety. The signing of the bond is to be witnessed and dated.

The significance of signing a contract as a deed is that a party is potentially liable for any breach of contract for 12 years. In this instance the surety's obligations are limited by the timetable identified in clause 6.

Schedule 7: Fluctuations Options

This Schedule provides details and procedures to assist in the assessment of price fluctuations which may occur during the construction of the Works. Reference should be made to clause 4.21, where it may be seen that this contract contains three optional fluctuation clauses, i.e. Option A, B and C.

Fluctuations Option A
Contribution, Levy and Tax Fluctuations

Fluctuations Option A is a limited fluctuations clause. It allows the Contract Sum to be adjusted in response to changes in taxes, levies, contributions, duties, etc. that are imposed by the government in respect of labour and materials. The following sections define the relevant taxes and levies and identify the labour and materials to which they may apply.

Deemed Calculation of Contract Sum – Labour

This section deals with changes in contributions, levies and taxes that are payable by an employer in respect of the labour he employs.

A.1   It is deemed that the Contract Sum will have been calculated in accordance with the details set out below, and it may be adjusted in response to the following events:

A.1.1   The Contract Sum will have been calculated taking into account the *'types and rates of contribution, levy and tax'* that are payable by the Contractor, as an employer, at the Base Date. For the purposes of this paragraph, the word employer is to be read in the general sense, i.e. a person or organisation who employs others under a contract of employment.

A.1.2   If, after the Base Date, there is a change in the amount payable by an employer with regard to the above tender rates or types, the value of the fluctuation will have to be assessed. The change may be the result of an increase or decrease in a tender rate, or the result of the introduction of or removal of a tender type (e.g. Selective Employment Tax was introduced by the government in 1966 and was subsequently phased out in 1973). Where a change has occurred, the net difference is to be calculated between the amount the employer (Contractor) would have paid, using the tender rates and types payable at Base Date, and what is actually paid when the new tender rates and types are applied. If the tender rate goes up (or a new tender type is introduced) the Contractor will be able to claim the net increase from the Employer, and if the tender rate decreases (or a tender type is abolished) the Contractor will have to credit the net amount to the Employer.

Any change in a tender rate that is payable under the Industrial Training Act 1982 is specifically excluded. For example, the Construction Industry Training Board charges many contractors an annual training levy; any changes in this levy are to be ignored for fluctuation purposes.

Labour fluctuations are to be calculated on the following categories of people:

A.1.2.1   workpeople (see definition paragraph A.11.3) who are on or adjacent to the site and are employed to work on the Works or in connection with the Works; and

A.1.2.2   workpeople who are directly employed by the Contractor (i.e. this excludes sub-contractors) who are not on or adjacent to the site but are producing materials or goods for the Works. This would cover the Contractor's operatives who may be working in a joinery workshop off-site. These people are eligible for fluctuations recovery for the period of time they are actually working on producing materials for the Works.

This final part of the clause confirms that the Contractor's fluctuations will be assessed by calculating the difference between what he would have paid if there had been no alteration in tender types or rates and what he actually pays as a result of a change in tender type or rate.

Example:

At base date, national insurance is payable by the Contractor at 10% of an operative's earnings above an earnings threshold of £80. Part way through the project, the national insurance contribution is changed to 12% with an earnings threshold of £90. An operative earns £450 in a week. The fluctuation for this one operative is:

Amount payable at base date rate:

$$(450 - 80) \times 10\% = £37$$

Amount payable after change in rate:

$$(450 - 90) \times 12\% = £43.20$$

Fluctuation:

$$43.20 - 37 = £6.20 \text{ payable to the contractor}$$

The Contractor is only allowed to claim for the net difference, there is no provision to claim profit and overheads on the fluctuation (but see paragraph A.12).

A.1.3   This clause details another category of employee for whom fluctuations may be calculated; that is, persons employed by the Contractor in connection with the Works, who are working on or adjacent to the site but do not fall into the category of 'workpeople' as defined in paragraph A.11.3. For example, this would include secretarial, administrative and supervisory staff working on site. To enable fluctuations to be calculated for these employees it is to be assumed that they are paid the same as a craftsman. The fluctuations may only be calculated where staff have been on site for two whole working days; part days are ignored (also see paragraph A.1.4.1).

A.1.4   This clause provides further information on how to assess the fluctuations allowed for in paragraph A.1.3. The details are as follows:

A.1.4.1   staff must have been on site for a minimum of two whole working days within the week. Part days cannot be added together to create a whole day. For example, a contracts manager is on site for the first half of every weekday excluding Friday, i.e. a total of 2 working days. However, this does not equate to 2 whole working days and therefore he could not be considered for fluctuations.

A.1.4.2   for the purposes of the fluctuation calculation, the staff are deemed to be paid the same rate as a craftsman employed by the Contractor (or sub-contractor in accordance with paragraph A.3) whose wage is set by the Construction Industry Joint Council (or other agreed wage-fixing body). Some wage-fixing bodies have agreements for various craftsman rates – depending on the type of craftsman and the skill level. Where the Contractor is employing craftsmen who have different agreed rates, then, for the operation of this paragraph, the highest craftsman's rate is to be used in the fluctuation calculation.

A.1.4.3   a definition of what is meant by 'employed by the Contractor' is that it covers employees who are covered by the PAYE Regulations 2003.

The following paragraphs A.1.5 to A.1.9 are very similar to some of the above paragraphs; the core difference is that these paragraphs relate to refunds or premiums that a Contractor may receive in his role as an employer.

A.1.5   It is deemed that the Contract Sum will have taken into account the types and rates of refund (receivable at Base Date) of the contributions, levies and taxes payable by the Contractor as an employer. The Contract Sum will also be deemed to have taken into account the types and rates of premium receivable (at Base Date) by the Contractor as an employer. For the purposes of the following paragraph, these allowances are referred to as 'tender type' and 'tender rate'.

A.1.6   If, after the Base Date, there is a change in the amount receivable by an employer with regard to the above tender rates or types, the value of the fluctuation will have to be assessed. The change may be the result of an increase or decrease in a tender rate, or it may be the result of the introduction of a new tender type, or the removal of a tender type. Where a change has occurred, the net difference is to be calculated between the amount the employer (Contractor) would have received, using the tender rates and types receivable at Base Date, and what is actually received when the revised tender rates and types are applied. If the tender rate goes up (or a new type is introduced) the Contractor will be able to claim the net increase from the Employer, and if the tender rate decreases (or a tender type is abolished) the Contractor will have to credit the net amount to the Employer.

A.1.7   This paragraph provides a definition of the word 'premium' as used in paragraphs A.1.5 and A.1.6. The term relates to a payment made (under an Act of Parliament) to a person who is acting as an employer, which alters the employer's cost of employing personnel.

A.1.8   This paragraph deals with the issue that some employees (i.e. workpeople defined in paragraphs A.1.2 and A.1.2.2) may be 'contracted out' under the Pensions Scheme Act 1993. Where such a situation exists, and for the purposes of assessing fluctuations, the Contractor's contributions will be calculated as though the employees were not contracted out.

A.1.9   This paragraph provides a definition of 'contributions, levies and taxes'. They are deemed to be all impositions (created by an Act of Parliament) placed on a person, acting as an employer, which affect the cost to an employer of employing personnel. It does not matter who is the recipient; it may be a government department or other body.

Deemed Calculation of Contract Sum – Materials

The following section sets out the rules and procedures for calculating fluctuations in relation to materials.

A.2   It is deemed that the Contract Sum will have been calculated in accordance with the details set out below, and it may be adjusted in response to the following events:

A.2.1   This paragraph sets out the basis upon which the Contract Sum is deemed to have been prepared. It is based upon the types and rates of duty and tax which are payable on '*the import, purchase, sale, appropriation, processing, use or disposal of the materials, goods, electricity, fuels, materials taken from the site as waste or any other solid, liquid or gas necessary for the execution of the Works*'. This provides a fairly broad definition of what may be considered as materials for the purpose of calculating fluctuations. The definition covers not only the materials used for the Works, but fuels and electricity, the disposal of unwanted materials and disposal of excavated waste and contaminants. The Contract Sum is based upon the types and rates of duty and taxes which are payable at the Base Date and have been introduced by an Act of Parliament. For the purposes of paragraph A.2.2 these types and rates are referred to as 'tender types' and 'tender rates'.

It is important to note that any VAT which the Contractor may claim as an input tax (i.e. the Contractor may recover his input tax from HMRC) is excluded from being treated as a fluctuation.

A.2.2 With reference to the above, if there is a change in tender type or tender rate which occurs after the Base Date, it will be necessary to assess the fluctuation. The fluctuation will be the net difference between what the Contractor now pays as a result of the change, compared to what he would have paid at Base Date.

Example:

At base date the fuel tax on diesel is 5%. During the progress of the project this is increased to 7%. The Contractor is entitled to claim the additional 2% on his diesel costs associated with this project.

Sub-let Work – Incorporation of Provisions to Like Effect

For most projects, the major portion of the Works will be carried out by sub-contractors. The following paragraph informs the Contractor of how he should incorporate appropriate fluctuation provisions into his agreements with sub-contractors.

A.3.1 Where a Contractor sub-lets part of the Works he is to ensure that the sub-contract agreement contains the same fluctuation provisions as Option A (with the obvious exclusion of paragraph A.3). The Contractor must also ensure that whatever percentage addition is allowed for in the Contract Particulars (see paragraph A.12) is also incorporated into the sub-contract agreement.

A.3.2 If the cost of the sub-contract works is increased or decreased, through the operation of the fluctuations clause in the sub-contract, then the net difference shall be payable to the Contractor or allowed to the Employer respectively.

Written Notice by Contractor

A.4.1 It is a requirement that the Contractor must notify the A/CA, in writing, when an event occurs in relation to one of the following paragraphs and which is relevant to this contract. The paragraphs are as follows:

A.4.1.1 paragraph A.1.2 – changes in tender rates and tender types in relation to contributions, levies and taxes payable on labour.

A.4.1.2 paragraph A.1.6 – changes in tender rates and tender types in relation to refunds and premiums receivable on labour.

A.4.1.3 paragraph A .2.2 – changes in tender rates and tender types in relation to duties and taxes payable on materials, etc.

A.4.1.4 paragraph A.3.2 – changes in cost of sub-contract sum as a result of sub-contract fluctuation.

A.4.2 Where the Contractor provides a written notice, in compliance with the above, it must be given within a reasonable time from the occurrence of the relevant event. The Contractor is not entitled to be paid fluctuation on an event unless he provides a written notice within a reasonable time, i.e. the written notice is a 'condition precedent' for payment.

Agreement – Quantity Surveyor and Contractor

A.5 For the purposes of this contract, the QS and Contractor may agree what they deem to be the net amount due to or from the Contractor as the result of a fluctuation event covered by Option A.

It is important to be aware of the limitations of this clause.

(See John Laing Construction Ltd v County and District Properties Ltd, 1982)

Fluctuations Added to or Deducted from the Contract Sum

The purpose of this paragraph is to explain how and when the fluctuation payments are to be reimbursed to the parties.

A.6 Any fluctuations that have been assessed in accordance with paragraphs A.1 to A.3 shall be dealt with by adding or deducting that amount from the following:

A.6.1 the Contract Sum; and

A.6.2 any amounts payable to the Contractor under clause 8.12.3.1. These amounts relate to payments made to the Contractor where his employment has been terminated following a default of the Employer, or as a result of the suspension of the Works under clause 8.11.

Any payments or deductions made under this paragraph are subject to the provisions and procedures contained within paragraphs A.7 to A.9.1.

Evidence and Computation by Contractor

A.7 To allow the cost of a fluctuation event to be properly assessed, the Contractor must provide whatever evidence and calculations the A/CA or QS may reasonably request. The Contractor is to respond to any request as soon as it is reasonably practicable. Where amounts are to be calculated under paragraph A.1.3 (i.e. fluctuations on non-workpeople) the evidence provided by the Contractor must include a signed certificate, confirming the authenticity of the information provided, for each week that a fluctuation is being calculated. For example, the Contractor may have been reasonably required to provide a weekly labour return (for his own staff and those of any relevant sub-contractors) identifying the non-workpeople for whom the fluctuation is being calculated and the times when they have been on site.

No Alteration to Contractor's Profit

A.8 There is to be no Contractor's profit added to the fluctuation sums assessed under paragraph A.6 (but see paragraph A.12).

Position Where Contractor in Default Over Completion

This paragraph deals with the situation where a Contractor, through his own default, has failed to complete the works by the completion date. In such a situation, an Employer may be concerned that the Contractor could now be affected by fluctuation events that occur after the completion date. In this case could the Contractor claim these costs back from the Employer? The answer is provided in the next paragraph.

A.9.1 The Contract Sum will not be adjusted as a result of any changes in tender rates and tender types occurring after the Completion Date. This is commonly referred to as 'freezing the fluctuations', and applies regardless of whether the tender rates or tender types are increased or decreased.

Example:

The completion date for a project is March 2009. At base date, an employer's national insurance contribution is 12.8%. During the course of the project, the national insurance contribution is raised to 13.5%, and a further increase to 14% is announced for April 2009. The Contractor is late in completing the Works, but his fluctuations claim for changes in national insurance is frozen at 13.5%; he cannot claim the later increase of 14%.

It is important to note that the operation of the above paragraph is dependent upon the provisions of paragraph A.9.2.

A.9.2 The above paragraph will not apply unless the following provisions are complied with:

A.9.2.1 the extension of time clauses (2.26 to 2.29) within the contract have not been altered or omitted; and,

A.9.2.2 where the A/CA has received a written delay notification from the Contractor (as clause 2.28), he has fixed or confirmed a Completion Date in writing which he considers to be correct in accordance with clause 2.28.

Work etc. to Which Paragraphs A.1 to A.3 Are Not Applicable

A.10 There are certain items which are not subject to Option A fluctuations, and they are as follows:

A.10.1 work which has been valued in accordance with clause 5.7, i.e. a daywork. A variation that has been valued as a daywork will have been priced at current rates and therefore must be excluded from any fluctuations calculation.

A.10.2 changes in the rate of VAT that is charged on the goods or services provided to the Employer by the Contractor. These changes will be accommodated within the VAT legislation. If an Employer is VAT registered, he may claim back the VAT he has paid. If an Employer is not VAT registered, he will have to pay whatever rate of VAT is in operation at the time of invoice.

Definitions for Use with Fluctuations Option A

A.11 The following definitions are provided as an aid to the drafting of the clauses in Option A.

    A.11.1 the Base Date (e.g. see paragraph A.1.1) is the date stated in the Contract Particulars (see Definitions and Contract Particulars, clause 1.1);

    A.11.2 materials and goods (see paragraph A.2.1) is to include timber that is used in formwork, therefore a Contractor could not claim for fluctuations in the cost of steel formwork. It is further pointed out that any other 'consumable stores, plant and machinery' do not fall within the definition of materials and goods;

    A.11.3 workpeople (see paragraph A.1.2) are people whose wages, etc. are determined by the Construction Industry Joint Council or any other relevant wage-fixing organisations within the construction industry, e.g. the Building and Allied Trades Joint Industrial Council;

    A.11.4 wage-fixing body (see paragraph A.1.4.2) means an organisation that sets down the terms and conditions that are recognised within that industry, and which will apply to the workers who come under its control;

    A.11.5 recognised terms and conditions (see above) are the workers' terms and conditions which have been agreed between the employers' organisations and the independent trade unions relevant to the industry. The agreement does not have to include all employer organisations and trade unions but it does need to include a substantial proportion.

Percentage Addition to Fluctuation Payments or Allowances

In July 1973 the JCT made changes to the current form of contract through the publication of Amendment sheet 8. This amendment introduced a provision whereby a percentage may be applied to the fluctuations calculated under Option A. The JCT never explained why they had introduced this percentage addition and did not provide any advice as to what percentage figure might be appropriate. Some commentators claim the percentage is an allowance for the Contractor to recover profit and overheads on the net fluctuations, and others claim it is an allowance to cover the Contractor's administrative costs in providing all the information and documentation necessary for the calculation of the fluctuation. If the Employer wishes a percentage addition to apply, the amount is to be inserted in the contract particulars. If this entry is ignored, there will be no percentage addition allowed.

A.12 Where allowed for in the Contract Particulars, a percentage addition shall be applied to the following:

    A.12.1 paragraph A.1.2 – changes in tender rates and tender types in relation to contributions, levies and taxes payable on labour,

    A.12.2 paragraph A.1.3 – fluctuations on staff who are not classified as workpeople,

    A.12.3 paragraph A.1.6 – changes in tender rates and tender types in relation to refunds and premiums receivable on labour,

    A.12.4 paragraph A.2.2 – changes in tender rates and tender types in relation to duties and taxes payable on materials, etc.

Fluctuations Option B

Labour and Materials Cost and Tax Fluctuations

Fluctuations Option B is a full fluctuations clause. It allows for changes in taxes and levies just the same as Option A, but it also allows for changes caused by market forces (e.g. inflation, supply and demand) in the cost of labour and materials. The following sections explain which labour and material may be considered for a fluctuations claim and how the claim should be assessed.

Deemed Calculation of Contract Sum – Labour Rates etc.

Paragraph B.1 deals with the cost of wages, expenses and the consequential effect on taxes and levies. Changes in taxes and levies are dealt with in paragraph B.2.

B.1   This paragraph identifies that the Contract Sum is deemed to have been calculated in accordance with the following details, and where it may be adjusted in response to the events set out below:

B.1.1   The Contract Sum is based upon the wage rates payable by the Contractor. The Contract Sum is also based upon other wage-related costs such as holiday pay, benefit schemes, employer's liability insurance and third party insurance. The wage rates, etc., are those payable by the Contractor on the following:

B.1.1.1   workpeople (see definition paragraph B.12.30) who are on or adjacent to the site and are employed to work on the Works or in connection with the Works; and

B.1.1.2   workpeople who are directly employed by the Contractor (i.e. this excludes sub-contractors) who are not on or adjacent to the site but are producing materials or goods for the Works. This would cover the Contractor's operatives that may be working in a joinery workshop off-site. These people are eligible for fluctuations recovery for the period of time they are actually working on producing materials for the Works.

The following sections set out the rules and conditions that will apply to the above workpeople:

B.1.1.3   the wage rates, etc. will be those set down by the rules and decisions of the Construction Industry Joint Council (or other applicable wage-fixing body) which have been promulgated (i.e. published) at Base Date;

It is important to understand the implication of the word 'promulgated' when used in this paragraph. It means that the Contractor is deemed to have made allowance in the Contract Sum for wage rates that have been agreed and published at Base Date. In some instances wage-fixing bodies have agreed annual wage increases spanning a three-year period. If an agreement has been promulgated at Base Date then the Contractor is expected to have included all the future increases within the Contract Sum;

B.1.1.4 the wage rates may include any incentive scheme or productivity agreement which complies with the rules of an applicable wage-fixing body; and

B.1.1.5 the wage rates for operatives regulated by the CIJC will be covered by the terms of the Building and Civil Engineering Annual and Public Holiday Agreements (or other agreements for operatives not covered by the CIJC). This agreement specifies how operatives in the construction industry will be paid for public and annual holidays. The Contractor is deemed to have taken into account the terms and agreements that have been promulgated at Base Date;

Finally it is stated that the Contract Sum is based upon any contribution, tax or levy which the Contractor is required to pay as an employer. The taxes, levies, etc. are to be calculated upon the basis of the wage rates and expenses identified above.

B.1.2 If the wage rates, etc. promulgated at Base Date are altered as a result of a change in the rules and decisions of a wage-fixing body, the net amount of the increase or decrease is to be calculated. It is also necessary to calculate other consequential costs, i.e. net increase or decrease in employer's liability insurance, third party insurance and contributions, taxes and levies payable by the Contractor on his labour costs.

B.1.3 This paragraph is virtually identical to the provisions contained within Option A, and relates to labour that does not fall within the definition of workpeople. See Option A paragraph A.1.3.

B.1.4 See Option A.1.4.

B.1.5 This paragraph provides more detail on the allowances that are deemed to be included in the Contract Sum. In the following sections information is provided on the travelling costs a Contractor may incur when employing operatives. The Contract Sum is based upon the following:

B.1.5.1 transport charges. In this instance the Contractor is entitled to submit a list of the transport charges he pays to his employees who are defined in paragraphs B.1.1.1 and B.1.1.2. The list is to be attached to the Contract Documents;

B.1.5.2 fares. These are the fares a Contractor pays to his employees (as paragraphs B.1.1.1 and B.1.1.2) in accordance with the rules and decisions of the appropriate wage-fixing body promulgated at Base Date.

B.1.6 This paragraph details how the fluctuation on transport charges and fares is to be assessed. The details are as follows:

B.1.6.1 if there is a change in the cost of transport charges compared to the list provided by the Contractor; or

B.1.6.2 if there is a change in fares as a result of an alteration to the rules and decisions of a wage-fixing body promulgated at Base Date, or if there has been an actual change in the cost of fares payable after Base Date;

then in response to any of the above events, the net amount of the increase or decrease is to be calculated, and the Contract Sum will be adjusted accordingly.

Deemed Calculation of Contract Sum – Labour Levies and Taxes

This paragraph deals with government changes in contributions, levies and taxes that are payable by an employer (in this case, the Contractor) in respect of the labour that he employs.

B.2 It is deemed that the Contract Sum will have been calculated in accordance with the details set out below, and it may be adjusted in response to the following events:

In many instances the wording of the following paragraphs are identical or similar to the paragraphs in Option A.

B.2.1 see A.1.1
B.2.2 see A.1.2
B.2.3 see A.1.3 and A.1.4
B.2.4 see A.1.5
B.2.5 see A.1.6
B.2.6 see A.1.7
B.2.7 see A.1.8, but refer to the additional information provided in paragraph B.2.7. Where an employee's employment is 'contracted out' for the purposes of an occupational pension scheme and in accordance with the rules of an appropriate wage-fixing body, any relevant fluctuations will be dealt with under paragraph B.1.
B.2.8 see A.1.9

Deemed Calculation of Contract Sum – Materials, Goods, Electricity and Fuels

B.3 It is deemed that the Contract Sum will have been calculated in accordance with the details set out below, and it may be adjusted in response to the following events:

B.3.1 the Contract Sum is based upon the market prices (i.e. the price the Contractor would normally expect to pay) that were current at Base Date and which apply to 'materials, goods, electricity, fuels or any other solid, liquid or gas' that are required to carry out the Works. The Contract Sum is also based upon the duty or tax that may be payable, at Base Date, for the disposal of site waste;
B.3.2 if, after Base Date, there is a change in the market price of materials or in the duty on waste disposal, the net increase or decrease is to be calculated and the Contract Sum is to be adjusted accordingly;
B.3.3 a change in market prices is to include any changes caused by any duty or tax imposed by an Act of Parliament on the 'import, sale, appropriation, processing, use or disposal' of any of the items identified in paragraph B.3.1. VAT that may be treated as an input tax by the Contractor is excluded from this paragraph. A Contractor is able to claim his VAT input tax from HMRC.

Sub-let Work – Incorporation of Provisions to Like Effect

B.4.1 If a Contractor sub-lets part of the Works, he is to ensure that the sub-contract agreement contains the same fluctuation provisions as Option B (with the obvious exclusion of paragraph B.4). The Contractor must also ensure that whatever percentage addition is allowed for in the Contract Particulars (see paragraph B.13) is also incorporated into the sub-contract agreement.

B.4.2 If the cost of the sub-contract works is increased or decreased, through the operation of the fluctuations clause in the sub-contract, then the net difference shall be added to or deducted from the Contract Sum accordingly.

Written Notice by Contractor

B.5.1 It is a requirement that the Contractor must notify the A/CA, in writing, when an event occurs in relation to one of the following paragraphs and which is relevant to this contract. The paragraphs are as follows:

B.5.1.1 paragraph B.1.2 – changes in wages and expenses;

B.5.1.2 paragraph B.1.6 – changes in transport charges or fares;

B.5.1.3 paragraph B .2.2 – changes in tender rates and tender types in relation to contributions, levies and taxes payable on labour;

B.5.1.4 paragraph B.2.5 – changes in tender rates and tender types in relation to refunds and premiums receivable on labour;

B.5.1.5 paragraph B.3.2 – changes in the market price or tax in relation to the cost of materials, etc.

B.5.1.6 paragraph B.4.2 – changes in cost of sub-contract sum as a result of sub-contract fluctuation.

B.5.2 Where the Contractor provides a written notice, in compliance with the above, it must be given within a reasonable time from the occurrence of the relevant event. The Contractor is not entitled to be paid fluctuation on an event unless he provides a written notice within a reasonable time, i.e. the written notice is a 'condition precedent' for payment.

The following paragraphs are virtually identical to those included in Option A.

Agreement – Quantity Surveyor and Contractor

B.6 see paragraph A.5 for an explanation

(See John Laing Construction Ltd v County and District Properties Ltd, 1982)

Fluctuations Added to or Deducted from the Contract Sum

B.7 see paragraph A.6 for an explanation

Evidence and Computations by Contractor

B.8 see paragraph A.7 for an explanation

No Alteration to Contractor's Profit

B.9 see paragraph A.8 for an explanation

Position Where Contractor Is In Default Over Completion

B.10   see paragraph A.9 for an explanation

Work etc. to Which Paragraphs B.1 to B.4 Are Not Applicable

B.11   see paragraph A.10 for an explanation

Definitions for Use with Fluctuations Option B

B.12   see paragraph A.11 for an explanation

Percentage Addition to Fluctuation Payments or Allowances

B.13   see paragraph A.12 for an explanation. Note the increased number of items which are subject to the percentage addition.

Fluctuations Option C

Formula Adjustment

Under Option C the contract operates on a full fluctuations basis similar to Option B; the major difference is that under Option C the fluctuations are calculated by using a series of published indices that are applied to an agreed formula.

Adjustment of Contract Sum – Formula Rules

C.1.1.1   The Contract Sum is to be adjusted in accordance with the procedures set out in Option C and in accordance with the Formula Rules issued by the JCT. The fluctuations will be assessed in accordance with the Formula Rules current at Base Date. A copy of the Formula Rules is available from the JCT website.

C.1.1.2   VAT is to be excluded from formula fluctuation calculations, and Option C has no effect on clause 4.6 (the VAT provisions within the contract).

C.1.2   The definitions that are to be found in the Formula Rules are to apply to the wording contained within Option C.

C.1.3   Formula fluctuation adjustments are to be taken into account in all payment certificates. For example, prior to the issue of any Interim Certificate it will be necessary to assess whether the value of work executed is subject to any formula fluctuation adjustment. Where the Employer is a local authority, the value of any formula fluctuation adjustment is reduced by the 'Non-Adjustable Element'.

Where applicable, the non-adjustable element is identified in the Contract Particulars and is stated as a percentage. For example, if the non-adjustable element is stated to be 15%, a Contractor will be entitled to receive only 85% of the fluctuations calculated under the Formula Rules.

C.1.4   Rule 5 of the Formula Rules identifies how and why previous fluctuation calculations may need to be corrected. When it is apparent that a correction is required, it is to be carried out and included in the next payment certificate to be issued.

Interim Valuations

C.2   Where Option C is used, it becomes a requirement that interim valuations must be carried out before the issue of each Interim Certificate (see clause 4.11).

The reason for this clause is that Option C fluctuations are assessed by applying a formula to the value of the work executed in the Interim Certificate, therefore it is important that the Interim Certificate is a reasonably accurate assessment of the work executed on site. Errors in the Interim Certificate will automatically lead to errors in the fluctuation assessment.

Fluctuations – Articles Manufactured Outside the UK

This paragraph relates to articles that are manufactured outside the UK, e.g. a lift car manufactured in Germany. It is not possible for the published indices to reflect changes in market prices in other countries or the possible impact of changes in currency values. Consequently, articles manufactured outside the UK are excluded from formula fluctuations and are dealt with as explained below:

C.3   This paragraph should be read in conjunction with rule 4(ii) of the Formula Rules. Where an article comes under this rule the Contractor should, at tender stage, provide a list identifying such articles. The list will identify the market cost in sterling (at Base Date) of the article delivered to site. The market price is deemed to include any duty or tax (excluding VAT) payable under an Act of Parliament. If, after Base Date, there is a change in the market price of an article, the net difference is to be calculated and the Contract Sum adjusted accordingly.

Power to Agree – Quantity Surveyor and Contractor

C.4   This paragraph identifies that the Quantity Surveyor and Contractor may, by agreement, alter the methods and procedures by which the formula fluctuations are normally assessed (i.e. in accordance with the Formula Rules). Where this occurs, it is deemed that the amounts calculated under such an agreement comply with the provisions of Fluctuations Option C, subject to the following:

C.4.1   an agreement to alter the methods and procedures will not be permitted unless the amount so calculated can be reasonably expected to be the same, or approximately the same, as if it were calculated in accordance with the Formula Rules;

C.4.2   any agreement under this paragraph must not alter the fluctuations that are payable by the Contractor to any sub-contractor.

Position Where Monthly Bulletins Are Delayed, etc.

Since January 2008 the Monthly Bulletins have been published electronically on the Building Cost Information Service web site.

C.5.1   If publication of the monthly bulletins is delayed or stopped, it will be impossible to accurately calculate the formula fluctuations. Where such a situation arises before the issue of the Final Certificate, the formula fluctuations are to be assessed on a fair and reasonable basis. For example, there may be other indices available that would provide an approximate indication of changes in building costs.

C.5.2   If the publication of the monthly bulletins resumes before the issue of the Final Certificate, any assessments that have been made on a 'fair and reasonable' basis are to be set aside and the adjustments are to be recalculated in accordance with Option C and the Formula Rules.

C.5.3   Where the publication of the monthly bulletins is delayed or stopped, the Contractor and Employer are still required to operate the procedures under Option C and the Formula Rules as far as they are able. The reason for this requirement is that, if the publication of the bulletins is resumed, it will be possible to quickly calculate the amount of formula adjustment due for the period when the indices were not initially available. For example, during this period interim valuations would be prepared in accordance with paragraph C.2, and formula adjustment proformas could be completed with the exception of the indices. Once the indices are published it would take very little time to complete the proformas and calculate the formula adjustment.

Formula Adjustment – Failure to Complete

C.6.1.1   Where the Contractor fails to complete the Works by the Completion Date, the formula adjustment will be 'frozen'. The value of work executed by the Contractor after the Completion Date will be adjusted using the indices for the valuation period in which the Completion Date falls.

Example:

Interim certificates are issued on the 12th of each month.

Contract completion date is 12th August.

The valuation period in which the completion date falls is 12th July to 12th August.

The mid-point of this valuation period is 28th July.

July indices will be used to assess the formula adjustment for this valuation period.

All future formula adjustments will be assessed using July indices.

C.6.1.2   This paragraph points out that if, for any reason, the above procedures are not followed, then the formula adjustments are to be 'corrected' to ensure that they do comply with paragraph C.6.1.1. For example, towards the end of a project there could be a degree of uncertainty as to what the Completion Date is as a result of requests for extensions of time and the omission of works. Consequently, once the A/CA has finalised the Completion Date, it may be necessary to alter previous formula adjustments.

C.6.2 The freezing of fluctuations as outlined above is only permissible if the following procedures are complied with:

    C.6.2.1 the extension of time clauses (2.26 to 2.29) within the contract have not been altered or omitted; and,

    C.6.2.2 where the A/CA has received a written delay notification from the Contractor (as clause 2.28) he has fixed or confirmed a Completion Date in writing which he considers to be correct in accordance with clause 2.28.

References

1. Defective Premises Act 1972, Chapter 35, Section 1(1), Office of Public Sector Information.
2. Standard Building Contract Guide (Sweet & Maxwell Limited, 2005).
3. Ibid.

PART 3

Legal Issues

There are a number of advantages to be gained from using standard forms of contract within the construction industry. Through continual use of these forms, members of the construction team become familiar with the format and content of the contract conditions and, as a consequence, have a better appreciation of their rights and obligations in relation to the administration of the associated construction project. The contract conditions of the standard forms are drafted by legal experts to ensure that the contract terms are clearly expressed and that they comply with current legislation and commercial practices. However, despite the use of these legal experts, there are occasions when disputes arise concerning the interpretation of certain words or phrases within the contract conditions, or there may be claims that certain terms are in conflict with current legislation, but this should not necessarily be viewed as a disadvantage. It is by such disputes that standard forms of contract have continued to evolve over the years. Where disputes have been settled through litigation, a record of the legal arguments put forward by the various parties and, more importantly, of the court's decision remains. As a result, an enormous volume of case law has developed over the years, which has helped to clarify how certain terms and conditions should be interpreted within the standard forms of contract. In certain instances, the court's decision has led to the redrafting of the contract or to the introduction of new contract conditions. To demonstrate this process, the following sections identify a number of areas where legal disputes have played an important role in clarifying the administrative procedures within the Standard Building Contract with Quantities (SBC/Q). In some instances, they have even resulted in the redrafting or introduction of new conditions within the contract.

Contract Particulars

The contract particulars is an important section of the SBC/Q. It is here that the employer and his advisers insert the information relating specifically to the employer's project. As a result, the contract particulars will provide much of the key material and information needed for the administration of the project. It is essential, therefore, that all sections of the contract particulars are completed correctly, otherwise the employer may face significant problems with his project, as demonstrated below.

Bramall & Ogden Ltd v Sheffield City Council (1983) was a case concerning the employer's right to deduct liquidated damages. Bramall & Ogden entered into a Joint Contracts Tribunal (JCT) Standard Building Contract (SBC) (1963 edition) with Sheffield City Council for the construction of 123 dwellings. The employer had inappropriately completed the liquidated damages section in the contract. The amount of liquidated damages that was to apply to the project was set 'at the rate of £20 per week for each uncompleted dwelling'. Although it is fairly clear what the employer intended, that is not how the liquidated damages should have been stated. The liquidated damages should have been based not on a single dwelling but on the cost to the employer of the whole site being incomplete, i.e. 123 dwellings $\times$ £20 = £2,460 per week. This error had significant implications for the employer.

As and when Bramall & Ogden completed a dwelling it was taken over by the council through the partial possession procedure. According to certificates issued by the architect, the project completion date was 4 May 1977 but the project was not finally completed until 29 November 1977. In a dispute referred to arbitration, the arbitrator awarded Sheffield City Council £26,150 in liquidated damages for the dwellings that were completed during the period of delay, i.e. between 4 May 1977 and 29 November 1977. Bramall & Ogden appealed against the arbitrator's assessment of the liquidated damages, as they argued that the partial possession procedure in the contract conditions required that the liquidated damages stated in the contract be reduced in proportion to the value of the work taken over by the employer through partial possession, i.e. if 75% of the works has been taken over by the employer then the liquidated damages figure is to be reduced by 75%. It was held that the council were not entitled to recover liquidated damages and that the arbitrator's award was incorrect. The judge considered that the way in which the liquidated damages had been set out in the contract was incompatible with the partial possession procedures and therefore unworkable, i.e. it was not possible to calculate a revised liquidated damages figure in response to the employer taking over dwellings through partial possession. However, although the judge decided that the council could not claim the liquidated damages as set out in the contract, he remitted the award to the arbitrator for reconsideration and commented that the council might be able to apply for unliquidated damages for breach of contract if the arbitrator were to give them leave to amend their pleadings. Unliquidated damages are damages that may be awarded through the arbitration process or by the courts. The party claiming the damages must prove the actual loss or damage suffered as a result of a breach of contract. The arbitrator or court will then decide the amount of damages, if any, that may be awarded.

In the previous case, the employer was fortunate in that he still had an opportunity to claim unliquidated damages from the defaulting contractor. However, in the case of Temloc Ltd v Erril Properties Ltd (1987), the employer was not so fortunate. In Temloc Ltd v Erril Properties Ltd (1987), the dispute related to the construction of a number of commercial units under a JCT Standard Building Contract without Quantities. According to certificates issued by the architect, Temloc were 47 days late in completing the works, a delay that resulted in Erril Properties claiming damages for late completion. However, when the contract documentation was prepared, the liquidated damages section had been inappropriately completed; the amount of liquidated damages had been inserted at '£nil'. Based on this evidence, Temloc argued that the contract documentation showed clearly that the level of liquidated damages agreed between the parties was £nil, i.e. liquidated damages were payable at zero pounds. Consequently, it was argued that, as the liquidated damages had been agreed at £nil Erril should not be allowed to submit a claim for unliquidated damages for the delay in completion. Erril Properties, on the other hand, claimed that the significance of inserting £nil in the contract documentation was to simply indicate that the parties had agreed that the liquidated damages clause would not apply to this project. If the interpretation put forward by Erril Properties were accepted, then they would still be entitled to submit a claim for unliquidated damages. Unfortunately the appeal judges did not accept the explanation given by Erril Properties. One of the judges commented that, elsewhere in the contract documentation where a clause was not to apply, it had been crossed out and initialled by the parties. It was held that the effect of inserting £nil in the contract was to confirm that there should be no liquidated damages for delayed completion by Temloc and, as a direct consequence, Erril were not entitled to submit an alternative claim for unliquidated damages.

In both of the above cases, the employer lost the right to claim liquidated damages through a failure to complete the contract particulars correctly. In both instances, it was stated that the employer could have achieved the effect they desired if they had amended the contract correctly.

Contract Bills Not to Override the Contract (Clause 1.3)

The contract bills will invariably contain a substantial amount of information concerning how a project is to be administered. On occasions, information contained in the contract bills may appear to conflict with, or affect, one or more of the express terms of the contract. To try and resolve this problem, the JCT has incorporated a condition into the contract to the effect that 'nothing contained in the Contract Bills…shall override or modify the Agreement or these Conditions'. The earlier 1963 version of the JCT contract had a virtually identical condition contained within clause 12(1) but it caused a few problems with regard to how it should be interpreted and applied. In the case of

English Industrial Estates Corporation v George Wimpey & Co Ltd (1972), a JCT contract had been entered into for the construction of an extension to a factory. In the bills of quantities, there was a provision for the employer to be allowed to place and install equipment during the progress of the works. There was a second provision that the employer be allowed to occupy and use any part of the works as soon as the employer (or architect) was of the opinion that this part of the works might be used or occupied without causing any delay to the works. As part of the dispute, questions arose as to whether these conditions in the bills of quantities contradicted the contract conditions relating to partial possession. If so, clause 12(1) of the contract conditions would render the provisions in the bills of quantities invalid. The judge in the first instance, and also the appeal judges, struggled to decide how clause 12(1) should be interpreted. There was a degree of uncertainty as to whether the contract bills could impose any obligation on the contractor or, alternatively, whether they could possibly override the contract. In the appeal, it was held that the provisions in the bills of quantities did not conflict with the contract conditions as they were there to explain the difference between occupation and use of the works by the employer, and partial possession. Two of the appeal judges were quite critical of the wording used within the contract and suggested that the contract should be revised to ensure greater clarity and consistency. A number of years later, the JCT eventually incorporated into the contract a provision allowing the employer to use or occupy part of the works (see clause 2.6).

Further confirmation of the difficulties the courts experienced in interpreting clause 12(1) of the 1963 JCT contract was provided by Moody v Ellis (1983). In a contract for the construction of a house, there was a requirement in the bill of quantities that the contractor was to prepare a programme, or statement, which set out clearly the sequence and timing of his work operations. In the JCT contract in use at that time, there was no requirement for the contractor to provide the architect with a construction programme. In the first hearing, the official referee held that the statement in the contract to the effect that the contract bills may not override or modify the contract meant that the contract bills could not impose an obligation on the contractor to provide a programme. This was overturned on appeal, where it was stated: 'There is however no question of this provision in the Bill of Quantities in any way overriding, modifying or affecting the conditions or any term of them; and we can see no reason why it should not take effect as part of the building contract.' The appeal court, therefore, supported the interpretation that the JCT had intended, i.e. that the contract bills may place further obligations on a contractor as long as they do not conflict with the contract terms. Perhaps in response to disputes like this, the JCT decided to incorporate a new clause into the contract (see clause 2.9.1.2) making it a contractual requirement that the contractor provides a programme.

Final Certificate

The final certificate, as its name implies, is the last certificate to be issued by the architect/contract administrator (A/CA). The purpose of the certificate is to bring closure to the project administration by confirming the final account figure for the project and that certain significant administrative tasks have been satisfactorily completed. The question is – how final is the final certificate? There is a mistaken belief among some contractors and sub-contractors that, once they have received their final certificate, their liabilities have ended. This is unlikely to be true unless the contract clearly states that the final certificate is conclusive evidence of satisfactory performance. It is, therefore, necessary to analyse the wording used by the JCT in order to identify whether or not the issue of the final certificate does bring an end to the contractor's liability.

When the 1980 edition of the SBC/Q was originally published, the contract draftsmen never intended the issue of the final certificate to signify the end of the contractor's liability but, because the wording they had used was open to different interpretations, this was the effect that they actually achieved. Two cases had been heard in the early 1990s[a], which questioned the effect of the final certificate, but the outcome of these cases provided conflicting advice. Fortunately, the issue was finally settled following the case of Crown Estate Commissioners v John Mowlem & Co. Ltd (1994). This was a case relating to the 1980 edition of the SBC/Q and concerned a dispute between the parties as to the meaning of the words used to define the conclusive nature of the final certificate. In the original version of SBC/Q 1980, clause 30.9 detailed what effect the issue of the final certificate would have on the parties. The JCT draftsmen had used the following wording in clause 30.9.1.1 to describe what effect the final certificate would have with reference to the works itself:

> *conclusive evidence that where and to the extent that the quality of materials or the standard of workmanship are to be to the reasonable satisfaction of the Architect the same are to such satisfaction…*

It was accepted by most construction professionals at that time that this clause referred only to a specific item of work, which the tender documents stipulated must be carried out to the architect's satisfaction. Most preamble clauses normally provide a clear description of the standard of workmanship required for the various items contained in the bill of quantities and, where the preambles bill provides no specification, the JCT contract specifies that the work shall be of a standard appropriate to the Works. However, an item may occasionally appear in the preamble section of a bill of quantities where the standard required is at the architect's discretion, e.g.:

Materials and Workmanship

Fibrous Plaster
Workmanship

All plaster coving and other decorative plasterwork is to be carried out to the Architect's satisfaction.

The problem posed by the example above is that, because there is no clearly defined or implied standard, the contractor will have no idea whether or not the decorative plasterwork has been executed satisfactorily until the architect expresses his approval

---

[a] Colbart Ltd v H. Kumar (1992), Darlington Borough Council v Wiltshier (Northern) Ltd (1993).

or disapproval, therefore the purpose of clause 30.9.1.1 was to signify the architect's acceptance of any such items; at least that was the intention of the JCT, although the eventual outcome was quite different, as demonstrated below.

J Mowlem were the contractors for a commercial development in Kensington. The final certificate was issued on 2 December 1992 even though there were some alleged defective works. More defects came to light after the issue of the final certificate and Crown Estate Commissioners gave notice of arbitration on 6 April 1993. This notice was more than 5 months after the issue of the final certificate and clearly outside the 28-day period (specified in the contract) within which the final certificate may be challenged. The defence put forward by J Mowlem was that, on their interpretation of the contract, clause 30.9.1.1 meant that the issue of the final certificate was conclusive evidence that the architect was satisfied with the works with reference to standards of materials and workmanship. Mowlem further pointed out that the client had not challenged the final certificate within 28 days from its issue; therefore, Mowlem could not be liable for any defective materials or workmanship.

On behalf of the Crown Estate Commissioners, it was argued that clause 30.9.1.1 should be viewed on a much narrower basis. Their interpretation was that the issue of the final certificate would only be conclusive for any item of work that had been identified in the contract documents as having to be 'executed to the architect's satisfaction'.

In the first hearing, the Crown Estate Commissioners were successful in having their interpretation accepted but, to the surprise of many, the Court of Appeal overturned the decision and supported the broader definition put forward by J Mowlem. The court considered that the architect's role was to review the whole works:

> ... the whole scheme of the contract meant that the architect must be satisfied as to the quality of all materials and the standard of all workmanship and form the opinion that they conform to those required by the contractual terms. For example, he must be so satisfied before he issues…the final certificate.

The court's interpretation was clearly not what the JCT had intended. The outcome of this case was that, once the final certificate has been issued, a contractor (and by implication all sub-contractors) cannot be held liable for any defects in materials or workmanship, the one proviso being that a client has the right to challenge the final certificate within 28 days of its issue.

The Court of Appeal's decision was delivered on 29 July 1994 and it was not until July 1995 that the JCT plugged this loophole with the publication of Amendment 15. This amendment redrafted clause 30.9.1.1 with the specific object of ensuring that the contractor would still be liable for defects in materials and workmanship even after the issue of the final certificate. The JCT explained the purpose of the amendment as follows:

> The amendments are intended to make clear that where the Contract Documents have not been complied with by the Contractor, and where the Employer takes such action as he may wish or be advised, then the issue of the Final Certificate will not be

a bar to any such action. Where, however, the Contract Documents make clear that the Employer and Contractor have agreed to abide by the decision of the Architect on certain requirements for the Works then the Final Certificate is conclusive evidence against both the Employer and Contractor that the Architect is satisfied that these requirements have been complied with.[2]

The current position regarding the issue of the final certificate is that it is conclusive evidence of the A/CA's reasonable satisfaction with reference to standards of materials or workmanship specifically identified in the contract documents as requiring to be carried out to the approval of the A/CA. The contractor, therefore, has no liability for such work after the issue and acceptance of the final certificate. In all other cases where the kinds and standards of work and materials have been described in the contract documents, the contractor will remain liable for such work after the issue of the final certificate. Where no standards of materials or workmanship are specified, the contractor is expected to provide standards that are appropriate for the works (clause 2.3.3), and again the contractor is liable for such work after the issue of the final certificate.

It is important to be aware that either party may challenge the issue of the final certificate through adjudication, arbitration or litigation. Any proceedings must be commenced within 28 days of the certificate being issued and, as a result, the final certificate will not be effective in respect of those matters that have been referred. It can be seen that either party has a relatively short period of time to decide whether or not to refer a dispute following the issue of the final certificate. It is therefore important to be aware of the day on which the 28-day period commences.

In the case of Cambs. Construction Ltd v Nottingham Consultants (1996), a question arose as to when the final certificate is actually issued, i.e. is it the date the architect prepares the certificate or is it the date when the certificate is received by the employer or contractor? In this dispute, the contractor was in disagreement with a client over payment for work carried out and had subsequently engaged a claims consultant (Nottingham Consultants) to assist in presenting the contractor's claim. No agreement had been reached with the client regarding the claim when the architect issued the final certificate, which was dated Tuesday, 13 October 1992. The obvious significance of this action was that, if the contractor did not dispute the final certificate within 28 days of its issue, then the certificate would be conclusive evidence that all loss and expense claims to which the contractor was entitled under the contract had been finally settled. The claims consultant was informed of the receipt of the final certificate but did not advise the contractor to serve a notice of arbitration until 11 November (29 days after the date on the certificate). Notice of arbitration was given to the employer on 12 November but was not accepted on the basis that it was too late, i.e. the employer claimed that the referral notice was not submitted within 28 days from the date of issue of the final certificate. Consequently, Cambs. Construction sued Nottingham Consultants for professional negligence as a result of their belated advice to challenge the issue of the final certificate. Counsel for the claims consultant argued that the date of issue of the final certificate was not on the day when the architect prepared and dated the certificate but the day when the certificate was actually received by the contractor, which in this case was claimed to be Monday, 19 October. On the basis of that argument, the notice of arbitration would have been within the specified 28-day period. Unfortunately for Nottingham Consultants the court did not support this proposal. It was held that the date of issue of a certificate was the day that it was posted. As a result of evidence produced it was accepted that the certificate in dispute was posted on either the 13 or 14 October, therefore the last day by which an effective notice of arbitration could have been given was either the 10 or 11 November. Following this judgement, a certificate is deemed to be issued on the day it is posted, which would normally be the day it is prepared and signed by the architect, or possibly the next day, depending upon the postal procedures adopted by the architectural practice concerned.

Fraud

Clause 1.10 of the SBC/Q identifies that, upon the issue and acceptance of the final certificate, certain matters may no longer be referred to adjudication, arbitration or litigation, unless there is evidence of fraud. For all other items where the final certificate does not provide conclusive evidence of satisfactory performance, the parties will remain liable for any breaches of contract for 6 or 12 years from the date of practical completion, depending upon whether the contract was entered into under hand or as a deed. This time period of 6 or 12 years during which the parties to a contract are liable is set out in the Limitations Act. However, if there is evidence of fraud, the guilty party may not be able to rely on this limitation period, as was demonstrated in Applegate v Moss (1979).

In this case, a developer had, in 1957, constructed a house that did not comply with the agreed specification; the house should have been constructed on a raft foundation but was instead constructed on poorly prepared strip foundations using concrete of an inadequate strength. The purchaser first became aware of this defect in 1965 when he tried to sell the property. He subsequently brought an action against the developer (Moss) for breach of contract, i.e. the failure to comply with the agreed specification. One of the defence arguments put forward by the developer was that the action was statute barred under the Limitations Act because more than 6 years had elapsed from when the cause of action arose. The developer's argument was not accepted by the trial judge as he considered that the defendant had been guilty of fraud within the meaning of the Limitations Act 1939. The relevant wording within Section 26 of the Act is set out below:

Where, in the case of any action for which a period of limitation is prescribed by this Act, either:

(a) *the action is based upon fraud of the defendant or his agents, or any person through whom he claims for his agent, or*
(b) *the right of action is concealed by the fraud of any person as aforesaid...the period of limitation shall not run until the plaintiff has discovered the fraud...or could with reasonable diligence have discovered it...*

The appeal judges supported the decision of the original trial judge and confirmed that the court's interpretation of the word 'fraud', in relation to the Limitations Act, was broader than the common law definition. In their view, fraud could be demonstrated by 'conduct which it would be "against conscience" for the defendant to rely upon'. As a consequence, it was held that the plaintiff's (Applegate's) right of action had been concealed by the covering up of the foundations and that, in accordance with the Limitations Act, this was fraud.

Note: the equivalent wording in Section 32 of the Limitations Act 1980 has been slightly altered although it should still convey the same meaning as the 1939 Act.

This view, that covering up defective work may constitute fraud, was confirmed in Gray v T P Bennett & Son (1987). In this case, bricklayers employed by a construction firm (McLaughlin) had hacked off the concrete nibs of the backing structure that were in place to support the brickwork. Some 14 years later, the brickwork started to bulge and the defective work was discovered. The brickwork had to be removed, the concrete nibs had to be reformed and the brickwork renewed. Could McLaughlin be held liable for this defective work after all this time? The judge said that the bricklayers deliberately concealed work, which they knew to be defective, which must amount to fraudulent concealment and that, in his view, the contractor's supervisory staff must have been aware of the concealment, therefore the contractor was still liable for the cost of the remedial works.

Choice of Materials

Clause 2.3 of the SBC/Q states that the contractor may not substitute any of the goods and materials described in the contract documents without the consent of the A/CA, and that the A/CA may not unreasonably withhold or delay giving his consent. Does this mean that, unless the A/CA makes a reasonable objection, the contractor has the right to substitute alternative materials from those specified? This question may be partially answered by Leedsford Ltd v The City of Bradford (1956), which was heard in the Court of Appeal. In this case, the bills of quantities specified that the artificial stone to be used on the project was to be obtained from the Empire Stone Company, or other approved firm. The contractor priced his tender using prices from two other stone companies, who provided much cheaper quotes than Empire. The architect would not give the contractor permission to use the cheaper stone and insisted that the stone must be purchased from Empire. The contractor claimed that this was a breach of contract, and sought to recover damages for the cost of the more expensive stone. The contractor claimed that the use of the phrase 'or other approved firm' was for the benefit of the contractor, so that, if he could obtain approval to use an alternative company who would supply artificial stone of a proper quality at a cheaper price, then he should be allowed to take advantage of that benefit. In consideration of that argument, Singleton LJ commented, 'I cannot help feeling that the words "or other approved firm" may well have been inserted into the bill of quantities to meet a position which might arise if the Empire Stone Company ceased business or if, perchance they were so busy that they could not supply the stone in time to bring about the completion of the contract within the period desired.' He went on to say that the words 'or other approved' did not give the contractor any extra rights as regards the selection of materials and that there was no obligation on the architect to approve the firms put forward by the contractor. The contractor's appeal was turned down. The current SBC/Q does not make use of the phrase 'or other approved firm' but the appeal court's decision may be useful in interpreting clause 2.3.

Deferment of Possession

Rapid Building Group Ltd v Ealing Family Housing Association Ltd (1984) was a dispute relating to a project for the construction of 101 dwellings comprising 5 blocks of flats. There were a number of disputes involved but one of the key issues was the late completion of the works by the contractor (Rapid) and the employer's (Ealing's) right to claim liquidated damages. The work was let on a JCT SBC and the contractor was to be given possession of the site on 23 June 1980. Unfortunately for Ealing, there were two squatters occupying part of the site, which meant that Rapid could not be given full possession of the site on the due date. The contractor was allowed full possession of the site approximately 19 days late and the architect issued an extension of time to cover this delay. However, the contract conditions at that time made no provision for the contractor to be given an extension of time where the employer had failed to give possession of the site. As a result, the architect's extension of time award was technically invalid. The breach of contract by the employer in failing to give the contractor possession of the site meant that 'time was at large', i.e. the contract completion date was no longer valid and the contractor's obligation was now to complete the works 'within a reasonable time'. As the contract completion date no longer existed, it was not possible for the architect to create a revised completion date by granting an extension of time. As a consequence, the employer was unable to claim liquidated damages. In the first hearing, the trial judge stated that, as a result of their breach, the employer had lost their rights to claim liquidated damages and that they could not now attempt to recover costs from the contractor through an action for unliquidated damages. The Court of Appeal agreed, for the most part, with the findings of the trial judge but differed on the question of unliquidated damages and held that Ealing were not prevented from pursuing a counterclaim for unliquidated damages.

This case illustrates the problems encountered by an employer when the contract conditions fail to meet their requirements. However, the JCT has now plugged this gap by offering employers an optional clause (clause 2.5) allowing them to defer the possession of site. Also, a further provision has been included in the contract conditions (clause 2.29.3) to enable the A/CA to award an extension of time in response to an employer's deferment of possession.

Contractor's Programme

Under clause 2.9.2, the contractor is obliged to provide the A/CA with 2 copies of his master programme, which is obviously an important administrative document as it outlines how the contractor intends to progress with the works, i.e. the order of operations and the time to be allowed to execute the works. However, by referring to the 'definitions' provided in clause 1.1, it is possible to see that the contractor's programme is not a contract document, therefore what obligations, if any, does the programme impose upon the parties to the contract? This was a question raised in the case of Glenlion Construction Ltd v Guinness Trust (1987). This was a project for the construction of a number of residential properties, to be completed within a 114-week period. The programme submitted by Glenlion indicated that they intended to complete the works within 104 weeks. It is normally accepted that a contractor is entitled to complete the works at a date earlier than the completion date given in the contract conditions. This fact is confirmed in clause 2.4, where it states 'the Contractor who shall…proceed with and complete the same (i.e. works) on or before the relevant Completion Date'. A number of disputes between Glenlion and Guinness had been referred to arbitration, and the ensuing court case was an appeal on certain questions of law. The appeal against the arbitrator's decisions was allowed to proceed because the judge considered that the questions of law were of importance to the construction industry since, at that time, there was little in the way of direct authority relevant to the issues raised. One of the questions raised was whether, according to the contract documents and the programme, Glenlion were entitled to complete their works by the programme completion date, i.e. before the contract completion date. Another question was whether or not there was an implied term in the agreement to the effect that the employer and his agents were to assist the contractor to carry out his works in accordance with the supplied programme so that he might complete the works by the completion date shown on the programme. The implication of this last question was whether or not the employer and architect were required to provide the contractor with all necessary instructions, drawings and details to allow the works to proceed at the accelerated rate indicated by the programme.

With regard to the first question, the judge confirmed that, in accordance with the contract conditions, Glenlion were entitled to complete the works earlier than the contract completion date, regardless of whether or not the programme showed an early completion date. As to the second question, the judge held that there was no implied term requiring the employer and his agents to assist the contractor to achieve the early completion date, but they may not deliberately obstruct the contractor's progress; their obligation is to work in such a manner that allows the contractor to complete the works by the contract completion date.

Notification of Discrepancies

Under clause 2.15, there is a requirement that, if the contractor finds any 'departure, error or inadequacy' or 'any other discrepancy or divergence' amongst the contract documents or the drawings and instructions issued by the A/CA, he should bring this to the attention of the A/CA. The contract conditions expressly require the contractor to give a notice only if he finds such a discrepancy, etc., but is there an implied requirement that the contractor should check through the documents to ensure that there are no conflicts or discrepancies? In the case of London Borough of Merton v Stanley Hugh Leach Ltd (1985), it was held that a contractor was under no such obligation. It was stated that there was an implied term that the architect would provide the contractor with accurate drawings and information and that 'the contract does not impose a duty on a contractor to check the drawings to see if there are discrepancies or divergences…'. Despite this case, questions have been asked as to whether a contractor has an implied duty to warn of any defect or failure in the design work of the architect. This question has been asked on a number of occasions and, although a number of answers have been provided, a degree of uncertainty still remains. The following cases illustrate the legal developments regarding a contractor's liability to warn the client of a defective design.

Equitable Debenture Assets Corporation Ltd v William Moss Group Ltd and Others (1984) is a typical example of what may happen when a project goes wrong. In this instance, problems arose with the sealing of curtain walling components, which resulted in water penetration. As frequently occurs in cases like this, it can be very difficult for the client to determine whether the fault lies in defective design or bad workmanship, or both. The end result was that the client sued all parties concerned, i.e. architect, consulting engineer, contractor and nominated sub-contractor. An important issue debated during this case was whether or not the contractor had a duty to warn the client of defective design. The judge formed the opinion that William Moss was a large and experienced construction company and, although they were required to construct the work as designed by the plaintiff's architect, it would be absurd for them to continue working once they became aware that part of the design would not work. The form of contract (JCT 63) did not require the contractor to check the architect's design; it only required the contractor to notify the architect of any divergences the contractor may find between the contract documents. Despite the lack of an express term it was held that, in this instance, the contractor owed a duty of care to the client and the architect (acting as an agent of the client) to inform the architect of design defects, which became known to him. It was consequently held that William Moss were in breach of an implied term to warn of the defective design and were also in breach of their duty of care in failing to warn the architect of the design defects upon construction. This was a judgement given in the first instance, and which never went to appeal.

The question of whether a contractor has a duty to warn the client of a defective design came before the courts once more. The case of Oxford University Press v John Stedman Design Group and Others (1990) was a complex multi-party dispute concerning defects in concrete floors and their granolithic finishings. As part of the proceedings, the architect (John Stedman Design Group) claimed that the contractor had a duty to warn the client of any deficiencies in the architect's design. The findings of the court were that the employer (Oxford University Press) had specifically employed an architect to provide designs for them and were not relying upon the contractor to advise them upon the suitability of any design. Furthermore, the judge could not see that there was any implied term in the contract between the employer and contractor (JCT 63 with 1977 Revision) indicating that the contractor should warn the employer of any design defect. The judge adopted the principle that design work was a matter of skilled judgement, upon which opinions may vary as to its suitability or otherwise. It would be impractical if contractors were expected to raise design concerns with the employer when they had no express contractual responsibility to do so and where the employer had employed an expert to provide the design. However, the judge did make one important proviso, which could have serious implications for contractors and sub-contractors: 'His Honour could not see any basis for the implication of a duty in tort owed by NH (the contractor) to warn of a design defect unless it might give rise to danger to the safety of persons or damage to some property other than that which was the subject of the defect.'[3]

The understanding at that time, therefore, appeared to be that there was no generally implied duty for a contractor or sub-contractor to warn of design defects unless it could be shown that there were specific terms in the contract requiring this duty, or that there was a special relationship between the parties and it was known that the plaintiff was relying upon the expertise of that party. However, there was also the proviso (made by HHJ Esyr Lewis in the above case) that liability may arise where a defective design causes danger to persons or other property.

The above point raised by HHJ Esyr Lewis was considered in the case of Plant Construction plc v Clive Adams Associates and JMH Construction Services (1999), which was heard in the Court of Appeal. Plant Construction, the main contractor, had contracted with the Ford Motor Company to carry out works at one of their research and engineering centres. Plant employed Clive Adams as consulting engineers for the project and entered into a subcontract with JMH. The contract entered into with JMH was based largely upon their quotation supplied to Plant and was for the execution of substructure work, including the design of temporary support work. During the progress of the work an engineer employed by Ford overrode the temporary support design put forward by JMH and gave instructions for a roof to be supported on Acrow props. The props failed to provide the necessary support and the roof collapsed. Fortunately no one was injured, as the works were unoccupied at the time of the roof collapse. Plant accepted that they were responsible to Ford for this failure and reached a settlement with them. Subsequently Plant sought to recover the cost of this settlement, plus their own costs in repairing the works (approximately £2 million), from JMH and Clive Adams Associates. Clive Adams settled with Plant and made a payment of £250,000. Plant continued their action against JMH and were successful in the first instance, although the court reduced the damages by 80% because of contributory negligence by Plant and Clive Adams. Despite having their damages reduced by 80%, JMH were still unhappy with the court's findings and appealed against the decision.

In the first hearing, it had been accepted, by HHJ Hicks, that the design variation issued by Ford's engineer was contractually binding on JMH and that consequently they could not be contractually liable to Plant for the design of the temporary works. This fact was not challenged in the Court of Appeal. What was pursued in the Court of Appeal was whether JMH had a duty to warn Plant about the deficiencies in the system proposed by Ford's engineer and how far they should have proceeded with such a defective system. It was held that JMH did have a duty to warn Plant about the design inadequacies of the temporary works, on the basis that JMH were under a duty to guard against the risk of personal injury to what was potentially a large number of people. In fact, this duty was considered to be an implied contractual duty of skill and care. Two important issues taken into account in reaching this decision were, firstly, that the proposed design of the temporary works was obviously dangerous and, secondly, that JMH knew it to be dangerous.

JMH had in fact told Plant's contracts engineer and consulting engineer of their concern about the design. They had even suggested an alternative approach but the consulting engineer dismissed this proposal. In the Court of Appeal, the question was asked whether JMH had done enough to discharge their duty to warn of design failings. The Court of Appeal held that JMH had not done enough to discharge this duty, although they did not suggest what actions JMH could have taken. Fortunately, the case was remitted back to HHJ Hicks, who did consider what actions JMH may have taken:

> In my judgement the safety element is central…a roof collapse endangered everyone beneath it.
>
> JMH should therefore have pressed its objections on this ground…objections… could and should have been progressively more formal and insistent if not met – for example by putting in writing if oral representations were ignored, by going to successively higher levels of management in Plant and Ford if lower levels did not respond – and they could have been accompanied by the threat or actuality of report to regulatory authorities. The crucial question is whether JMH could and should, in the last resort, have refused to continue work if the safety of workmen was at risk…I am clear that it could and should have done so.

The Court of Appeal reserved for future consideration the circumstances of what may happen where:

- a contractor did not know that a design was dangerous, although it was arguable that he should have known; or
- a contractor knew about a design defect, or ought to have known about it, but the defect was not dangerous.

From the above it would appear that the contractor does not have an obligation to check through all the contract documentation to find any discrepancies, errors or omissions but, if the contractor is aware of a potentially dangerous design problem, he must inform the designer and/or the employer of any potential defects in the design details, although such advice may not always be well received. Where a design defect is potentially dangerous and warnings go unheeded, it may be necessary to suspend carrying out the work.

Unfixed Materials and Goods

It is a condition under the SBC/Q that the employer will normally pay the contractor for materials that have been delivered to site, and may in certain circumstances pay for materials that are being stored off site. Within the contract conditions, it states that, where the employer has paid for such materials, they become his property, but can this be overridden where a supplier has made use of a 'retention of title' clause in his contract of sale with the contractor? This is a question that caused considerable concern for many client organisations during the latter part of the twentieth century. During this period, many suppliers attempted to protect themselves from the insolvency of their customers by including in their contracts of sale express conditions, which prevent the title in the goods from passing to the customer until the goods have been paid for (a procedure that is allowed under Section 19, Sale of Goods Act). The following provides an example of terms typically used by suppliers to try and protect their interests:

> The property in the goods supplied shall not pass to the Purchaser until the Company has received full payment of the contract price thereof and payment of all further sums at the time of delivery of the goods owing by the Purchaser to the Company, and the Company may enter on to the Purchaser's premises or any premises in the Purchaser's occupation or control (whether exclusive or not) in order to repossess goods for which the contract price has not been fully paid.

> If the Purchaser shall sell the goods or shall incorporate them in or with any other product or products before payment of the contract price to the Company, he will be deemed to hold the proceeds of such sale on trust for the Company and must account forthwith to the Company for the same and in the event of the goods supplied being incorporated in or with other products the product thereof shall become and or shall be deemed to be the sole and exclusive property of the Company.

The above conditions relating to ownership of materials have been drawn up to protect the seller where a buyer becomes insolvent, the intention being that the seller can reclaim his goods where they are still in the possession of the buyer, or claim a right in products into which the goods have been incorporated, or claim from the proceeds of sale where the goods/products have been sold on to a third party. Without a retention of title clause, the only remedy available to the seller would be to sue for the monies due but, if the buyer is insolvent, the debt will be ranked alongside all the other unsecured creditors, with little likelihood of any payment being received.

For some commercial organisations the inclusion of a retention of title clause in their contracts of sale has been common practice for many years, but it was not until a dispute arose between Aluminium Industrie Vaassen BV and Romalpa Aluminium Ltd in 1976 that the implications of such a clause reverberated around the construction industry. Aluminium Industrie Vaassen (AIV) were a Dutch company, which had sold aluminium foil to Romalpa. In their conditions of sale, AIV had stated that:

- the title in the foil did not pass to Romalpa until all monies due had been paid in full;
- if the materials were incorporated by Romalpa into any goods, AIV was able to claim a title in such goods.

As a consequence of financial difficulties, Romalpa had a receiver appointed by a bank, which possessed a floating charge over the company's assets. At that time, AIV were owed £122,000 and, through the terms in their contract of sale, they claimed the goods that remained on the premises of Romalpa and presented a claim for the value of the material sold on by the receiver. AIV were successful in their claim and the Court of Appeal held that AIV were entitled to the goods that remained in the possession of Romalpa; the receiver was also required to hand over the £35,152 received from the sale of aluminium foil to subpurchasers. It was deemed that Romalpa held the proceeds in a fiduciary capacity for AIV, i.e. Romalpa were held to be bailees of the foil and were accountable to AIV for the proceeds. The subpurchasers were entitled to retain the materials as they had bought them in good faith from a 'buyer in possession'. This principle of a 'buyer in possession' is an important issue for employers, as explained below.

As a result of their retention of title clause, AIV were able to gain priority over preferential creditors, and secure payment or possession of materials they had supplied. The decision in this case received widespread coverage from the construction media, resulting in many organisations inserting 'Romalpa-type' clauses in their contracts and causing great concern among client organisations who became very reluctant to pay for materials on site. In hindsight, there was no need for clients to be so concerned, as part of the judgement given in the Romalpa case stated that a subpurchaser (i.e. an employer) would not be affected by a retention of title clause where they had purchased the goods from a buyer in possession (i.e. a contractor) and had no prior knowledge of the existence of the retention of title clause. This principle of a buyer in possession being able to pass a good title to a subsequent purchaser is confirmed within Section 25 of the Sale of Goods Act:

> *Where a person, having bought or agreed to buy goods obtains, with the consent of the seller, possession of the goods or the documents of title to the goods, the delivery or transfer by that person, or by a mercantile agent acting for him, of the goods or documents of title, under any sale, pledge, or other disposition thereof, to any person receiving the same in good faith and without notice of any lien or other right of the original seller in respect of the goods, shall have the same effect as if the person making the delivery were a mercantile agent in possession of the goods or documents of title with the consent of the owner.*

Therefore, as long as a subpurchaser is unaware of any lien on the property he should obtain a good title, even though the party he is buying from does not have good title. There is normally no obligation on the subpurchaser to ascertain whether there are any restrictions on the seller's rights to sell the goods. To understand how a 'Romalpa clause' and the Sale of Goods Act may affect the parties to a construction project, it is worthwhile considering the following construction-related disputes.

In W Hanson (Harrow) Ltd v Rapid Civil Engineering Ltd and Usborne Developments Ltd (1987), Usborne were involved in the development of three sites in London and had engaged Rapid to carry out the construction work. Hanson were timber suppliers who had agreed to supply Rapid in accordance with Hanson's standard conditions of trading. Rapid were experiencing financial difficulties and on 15 August 1984 had a receiver appointed, but Hanson, who at that time were unaware of the receivership, made a delivery of materials to one of the sites on the following day. On learning of the receivership, Hanson attempted to reclaim their materials from Rapid and relied upon their standard conditions, i.e.:

(a) The property in the goods shall not pass to you until payment in full of the price to us.
(b) The above conditions may be waived at our discretion, where goods or any part of them have been incorporated in building or construction work.

The receivers refused to make payment for materials supplied and would not allow Hanson to reclaim their materials, although they did allow Hanson access to site to enable them to prepare an inventory of the unpaid materials. Hanson then approached Usborne (the developer) to inform them of their retention of title clause and to seek assurances that Usborne would not attempt to make use of the disputed materials. Not surprisingly, perhaps, Usborne would give no such assurance, and Hanson commenced proceedings to claim the value of the goods, £6,768.97.

One of the issues was whether or not Usborne had acquired a good title in the materials. It was claimed by Usborne that they had acquired the title under Section 25 of the Sale of Goods Act, i.e. a buyer in possession (Rapid) may pass a good title to a subsequent purchaser (Usborne). In this instance, however, Usborne had not made payment to Rapid for the valuations that included the disputed materials, and within the main contract there was a provision that the property in goods supplied to the sites would not pass to the employer until the valuation payment had been made.

Usborne had not complied with the condition that required payment through the valuation procedure before there could be a transfer of property from Rapid; therefore, although there had been an agreement to sell, there had not been a sale and Usborne could not claim title under Section 25 of the Sale of Goods Act 1979. Consequently, Hanson were entitled to exercise their rights under the retention of title clause and reclaim materials on site that had not been paid for by Usborne. Hanson would not have been able to claim the materials back if Usborne had made a payment to Rapid, unless Usborne had been given prior notice of the retention of title clause. Hanson's claim would also have been defeated where any of the materials had been incorporated into the works, as the materials would then be attached to the property.

In Archivent Sales Developments Ltd v Strathclyde Regional Council (1984), Archivent supplied ventilators to a contractor (R D Robertson) involved in the construction of a school for Strathclyde. The works were based on the 1963 edition of the JCT SBC/Q (73 Revision). Archivent had the following conditions in their contract of sale:

> Until payment of the price in full is received by the company, the property in the goods supplied by the company shall not pass to the customer.

The ventilators were delivered to site and included for payment in interim certificate number 13; the employer paid Robertson for the materials but the contractor failed to make a payment to Archivent. The contractor subsequently had a receiver appointed by the Clydesdale Bank, which meant that Archivent were unlikely to receive payment for their goods. Archivent consequently contacted Strathclyde for the return of their ventilators (relying on their retention of title clause) or, alternatively, their value. Strathclyde declined in both instances and relied upon Section 25 of the Sale of Goods Act to argue that they had no knowledge of any retention of title and had acquired the ventilators in good faith, thus granting them good title. Archivent argued that the contract (JCT 63) did not make any provision for the goods to be transferred from the contractor to the employer (i.e. the goods remained in the control of the contractor; there was no transfer of property, therefore the Sale of Goods Act was not applicable). The court held that the goods were transferred from contractor to employer when the surveyor incorporated the goods into his interim valuation and the employer subsequently made payment to the contractor. Although Archivent's retention of title was properly constructed and valid so that the contractor did not have a title in the goods, the employer still gained title through Section 25 of the Sale of Goods Act. The important point was that the employer had acted in good faith, having received no prior notification of any lien on the goods.

Although these cases relate to older versions of the JCT contracts, the same outcome would be achieved where the current SBC/Q is being used, therefore it would appear that employers have little to fear when they pay contractors for materials that are stored on or off site. However, the above cases related to situations where the employer paid the contractor for materials purchased by the contractor. On many projects, these days the majority of work is carried out by sub-contractors and, as a result, a large proportion of materials on site may have been purchased by sub-contractors, and not by the contractor. This can cause a potential problem for employers because a subpurchaser (i.e. employer) may acquire a good title in the goods and materials only when the Sale of Goods Act is applicable; where a sub-contractor supplies materials to site, this will be covered by a contract for the supply of work and materials, not by the Sale of Goods Act and, as a result, an employer would not be able to claim a good title under Section 25 of the Sale of Goods Act. This potential problem was highlighted in the case of Dawber Williamson v Humberside County Council (1979). This case related to a school being built for Humberside by main contractors Taylor & Coulbeck Ltd. The contract documentation was based predominantly upon JCT 63 (a forerunner of the current SBC/Q).

Dawber Williamson were domestic sub-contractors employed by Taylor & Coulbeck Ltd. In November 1976, Dawber Williamson had delivered to site 16 tonnes of roofing slates, which were subsequently included in a valuation as materials on site. The value of the slates was included in the subsequent interim certificate and the monies paid to Taylor & Coulbeck by Humberside. However, the contractor failed to pass the monies on to Dawber Williamson and, shortly afterwards (24 January 1977) Taylor & Coulbeck went into liquidation. Dawber Williamson went to the site and attempted to reclaim their slates but were refused access by the employer. After negotiations, Dawber Williamson eventually completed the roofing work, but on a labour-only basis. On 20 May 1977 they served a writ on the employer for the value of the slates, £3,754.62 plus damages. In their defence of the action, Humberside relied upon the contract conditions between themselves and the contractor and the subcontract conditions between the contractor and Dawber Williamson (i.e. JCT 63 and 'Blue Form of Sub-Contract') as follows:

> *…any unfixed materials and goods delivered to, placed on or adjacent to the Works…shall not be removed…unless the Architect has consented in writing…*

JCT 63, Clause 14 (1)

> *Where the value of such materials or goods has…been included in any Interim Certificate under which the Contractor has received payment, such materials and goods shall become the property of the Employer.*

JCT 63, Clause 14 (2)

> *The Sub-Contractor shall be deemed to have notice of all provisions of the Main Contract…*

National Federation of Building Trades Employers (NFBTE) Blue Form
Clause 1(1)

At first glance, it would appear that the employer had a cast iron case based upon the quoted contract conditions. It was argued that clause 1(1) of the subcontract notified the sub-contractor that, where materials had been brought to site and paid for through an interim certificate, those materials were to become the property of the employer. However, the Court did not support that view. Their decision was that:

- Clause 1(1) of the subcontract did not operate to the extent that clause 14(1) of the main contract should be viewed as part of the subcontract conditions.
- Clause 1(1) did not create privity of contract between the employer and sub-contractor, therefore the employer was unable to enforce clause 14(1) of the main contract against the sub-contractor.
- Clause 14(1) could work effectively for the client only where the main contractor owned the materials or was authorised by the sub-contractor to act as their agents in disposing of their title in the goods. The subcontract made no provision for the title in the goods to pass from the sub-contractor to the contractor, therefore the contractor could not pass on a good title to the employer under clause 14(1).

The implication of this judgement was that the contractor never owned the slates and therefore could not pass ownership on to the employer; the sub-contractor was not party to the main contract, and could not be bound by its conditions. The end result was that Humberside had to pay Dawber Williamson the value of the slates plus 10% interest from the end of January to the date of the writ; therefore, despite the apparent protection provided to the employer through the terms contained in the main contract, Humberside had to pay once more for materials, which they believed they already owned.

As a consequence of this case, the JCT introduced a new condition into the SBC/Q requiring the contractor to include the following terms into any subcontract entered into:

> where, in accordance with clauses 4.10 and 4.16 of these Conditions, the value of any such materials or goods has been included in any Interim Certificate under which the amount properly due to the Contractor has been paid to him, those materials or goods shall be and become the property of the Employer and the sub-contractor shall not deny that they are and have become the property of the Employer;

SBC/Q, clause 3.9.2.1

> if the Contractor pays the sub-contractor for any such materials or goods before their value is included in any Interim Certificate, such materials or goods shall upon such payment by the Contractor be and become the property of the Contractor.

SBC/Q, clause 3.9.2.2

The purpose of the above terms is to provide a means by which the title in the sub-contractor's materials may pass to the contractor. However, it is not certain how effective these terms are in protecting the employer's interests; according to one legal authority, the amendments are unworkable[4] since the employer is not party to the subcontract and is therefore unable to enforce the conditions in the subcontract. It is for situations like this that the Contracts (Rights of Third Parties) Act 1999 was introduced.

Adjustment of Completion Date

Owing to the ambiguity of the wording used in the 1963 edition of the SBC/Q, it was unclear if it was a condition precedent that the contractor must give a notice of delay before the architect is required to consider whether or not an extension of time should be awarded. In the case of London Borough of Merton v Hugh Stanley Leach Ltd (1985), the Court of Appeal carefully analysed the relevant wording in the contract and made reference to Keating.[5] They came to the conclusion that the architect was under a duty to both the employer and contractor to review the completion date once he was aware of an event that entitled the contractor to an extension of time. It was stated that the architect could not 'ignore events which he knows are likely to cause delay beyond completion date even though, to the knowledge of the architect, the contractor is not aware that the progress of the works is delayed'. As a consequence, if an architect should refuse to consider an extension of time, on the grounds that the contractor had failed to provide a notice of delay, this could result in the employer losing his right to deduct liquidated damages. A further ambiguity in the contract wording meant that it was also unclear whether a contractor is to provide notice not only of a delay that has actually occurred, but also any delays that are likely to arise in the future. The court's decision was that a contractor was not required to provide notice of a delay that might arise from an event in the future, but a notice should be provided where an event has already occurred, which will cause a delay in the future. Finally, the court considered the implications where a contractor fails to give a notice of delay, or provides a late notice, so that the architect is unaware of the potential delay to the works. Again the court quoted from Keating, to the effect that in such circumstances the architect can: 'take into account that the contractor was in breach of contract and must not benefit from this breach by receiving a greater extension than he would have received had the architect, upon notice at the proper time, been able to avoid or reduce the delay by some instruction or reasonable requirement'.

The wording of the current SBC/Q has been modified since the above case. It is now clear that, prior to practical completion, an extension of time will be considered only upon receipt of a contractor's notice. However, once practical completion has been achieved, the A/CA is under an obligation to review the completion date, taking into account relevant events that may not have been notified by the contractor. The SBC/Q now clearly requires the contractor to provide a notice when it becomes reasonably apparent that the works is likely to be delayed, i.e. by a future occurrence. Finally, contractors should be aware of the above comments by Keating and ensure that any notices regarding a delay to the works are promptly forwarded to the A/CA.

Practical Completion

The issue of the practical completion certificate is an important event during the progress of a project. The decision as to whether a project has achieved practical completion is a matter of opinion for the A/CA, but how does the A/CA or contractor decide that practical completion has been achieved? The contract conditions do not provide a definition of the term 'practical completion' but there have been a number of cases where the courts have attempted to provide a definition. In City of Westminster v J Jarvis (1970), a case which was eventually referred to the House of Lords, Viscount Dilthorne made the following statement:

> The contract does not define what is meant by 'practically completed'. One would normally say that a task was practically completed when it was almost but not entirely finished; but 'Practical Completion' suggests that that is not the intended meaning and that what is meant is the completion of all construction work that has to be done.

The wording in the JCT contract at that time made use of the phrases 'practically completed' and 'practical completion'. The current SBC/Q refers only to 'practical completion' and, as can be seen above, Viscount Dilthorne was of the opinion that this means that all construction work should be completed and there should be no patent defects.

The meaning of the term 'practically complete' was considered by the courts again in the case of H W Nevill (Sunblest) Ltd v Wm Press & Sons Ltd (1981), where Judge Newey commented as follows:

> I think that the word 'practically'...gave the architect a discretion to certify that William Press had fulfilled its obligations..., where very de minimis work had not been carried out, but that if there were any patent defects in what William Press had done the architect could not have given a certificate of practical completion.

Judge Newey expanded on his views relating to practical completion in Emson Eastern Ltd (In Receivership) v EME Developments Ltd (1991), where he said: 'building construction is not like the manufacture of goods in a factory. The size of the project, site conditions, use of many materials and employment of various types of operatives make it virtually impossible to achieve the same degree of perfection as can a manufacturer. It must be a rare new building in which every screw and every brush of paint is absolutely correct.'

It should be noted that the above opinions expressed by the courts are not legally binding and that the current SBC/Q makes no mention of the work being practically complete. However, it would appear that there is strong support for the approach that practical completion is achieved when the works are completed with the exception of minor defects. An A/CA would be advised not to issue a practical completion certificate where there are significant patent defects, because the issuing of the certificate might imply that the defects have been accepted and do not need to be remedied (but see Pearce & High v Baxter).

Non-completion Certificate

Under clause 2.31 of the SBC/Q, if the contractor fails to complete the works by the completion date, the A/CA should issue a non-completion certificate. This certificate, along with a notice from the employer, must be issued before the employer is entitled to deduct liquidated damages. In an earlier edition of the SBC/Q (JCT 1980), there was some ambiguity concerning the issue of the non-completion certificate. The wording in the contract at that time implied that a non-completion certificate should be issued where a contractor is late in completing the works. The employer is to notify the contractor of the amount of liquidated damages required from the contractor and then the employer may deduct the liquidated damages. It was also stated that, if the architect were to subsequently fix a later completion date, the employer would have to repay any liquidated damages for the period up to the later completion date. The interpretation of these conditions was disputed in the case of A Bell & Son Ltd v CBF Residential Care and Housing Association (1989). Bell were late in completing the works and therefore the architect issued a non-completion certificate and the employer gave notice of their intention to deduct liquidated damages. The architect subsequently issued three more extensions of time, culminating in a revised completion date of 20 May 1986. Practical completion was achieved on 18 July 1986 and the final certificate was issued on 25 February 1988. The employer paid the final certificate less the liquidated damages calculated from 20 May to 18 July. The contractor disputed the employer's right to deduct the liquidated damages on the basis that the employer and architect had failed to follow the proper procedures. The decision of the court was that an architect's non-completion certificate becomes invalid where he subsequently grants a further extension of time and he must therefore issue a new non-completion certificate if the contractor fails to meet the newly revised completion date. The court adopted the same approach with reference to the employer's notice and stated that, in the above circumstances, the employer must issue a new notice before he may deduct liquidated damages. The Court of Appeal came to a similar decision in Finnegan (JF) v Community Housing Association (1993). As a result, changes have been made to the wording of the SBC/Q so that it now states clearly that the fixing of a new completion date cancels a previously issued non-completion certificate and that a new certificate is to be issued where necessary. The wording for the employer's notice has been changed to the effect that, once a notice has been issued, it will now remain in force despite any further extensions of time being granted. However, if an employer should wish to deduct liquidated damages from payments due to the contractor, he must first issue the appropriate withholding notice to ensure compliance with the Housing Grants, Construction and Regeneration Act (HGCRA) (clause 4.13.4).

Rectification Period

The Rectification Period (referred to in earlier JCT contracts as the defects liability period) will follow on from the date of practical completion. During the Rectification Period, the A/CA may instruct the contractor to remedy defects that have come to light and he may also issue the contractor with a schedule of defects. The latest date by which the schedule of defects is to be issued is no later than 14 days from the end of the Rectification Period. The A/CA may issue only one schedule of defects, and after its issue he may no longer issue instructions to the contractor to remedy any further defective work. Consequently, most A/CAs issue the schedule of defects towards the end of the Rectification Period, or just after. Once the contractor has received the schedule of defects, can he still be liable to the employer for patent defects that had not been notified? This issue was raised in Pearce & High Ltd v Baxter and Anor (1999). The works comprised alterations and an extension to a house and was let on the JCT Minor Works contract. The practical completion certificate was issued on 13 November 1995 and the defects liability period expired on 13 May 1996. On 29 January 1996, the contractor (Pearce & High) commenced proceedings in the County Court to recover monies they claimed were outstanding. The employer (Baxter) submitted an initial defence on 20 February but then submitted a further defence and counterclaim on 8 October 1996, in which they claimed for the cost of remedying alleged defects, which became evident during the defects liability period, although the defects had not been notified to the contractor during that time. The trial judge held the view that Baxter had lost the right to claim damages for the alleged defects because they had failed to notify the contractor in due time. He stated the following: 'if a building owner does not notify defects within the defect liability period then…the contractor having been denied the opportunity of returning to the building, cannot thereafter be sued in respect of patent defects which are not notified to him.' This decision was overturned on appeal when it was stated that, where the contractor had carried out defective work, this was a breach of contract and the procedures relating to the defects liability period do not prevent the employer from recovering damages at a later date. The earlier case of William Tomkinson and Sons Ltd v The Parochial Church Council of St Michael (1990) was referred to in support of this view. However, the employer was not entitled to claim the full cost of having the defects remedied. The appeal judge commented that the defects liability period has advantages for both parties; it entitles the employer to request the contractor to return to site and remedy any defects and it gives the contractor the right to return to site to remedy any defects that are notified. A contractor will normally be able to rectify defects in his own work far more cheaply than if the employer were to engage a third party to carry out the work. As a consequence, because the contractor was not notified of the patent defects during the defects liability period, or through the issue of the schedule of defects, the employer was not entitled to recover the full cost of the remedial work. 'The employer cannot recover more than the amount which it would have cost the contractor himself to remedy the defects.'

Replacement of Architect/Contract Administrator

In the case of Croudace Ltd v The London Borough of Lambeth (1986), the contractor (Croudace) was in the process of trying to agree a loss and expense claim when Lambeth's chief architect retired and was not replaced. Subsequently, there was some dispute as to whether there was an architect in place to instruct the ascertainment of the loss and expense. The judge in the first instance stated that the contract conditions gave Lambeth the power to nominate a replacement architect and, furthermore, that they had an implied obligation to renominate. This was supported in the Court of Appeal, where it was stated that Lambeth's failure to nominate a successor to the retired chief architect was a breach of contract. The current SBC/Q contains a procedure for the nomination of a replacement A/CA; it states clearly that the employer is obliged to renominate where necessary and provides a time frame by when it must be achieved.

Workmanlike Manner

Under clause 2.1, the contractor is required to carry out the works in a proper and workmanlike manner. In earlier versions of the SBC/Q there was no mention of this. However, this was changed in response to the case of Greater Nottingham Co-operative Society Ltd v Cementation and Others (1988). Although this was a dispute between an employer and a nominated sub-contractor, the outcome of the case had potential implications for the relationship between the employer and contractor. The basic facts of the case are as follows.

Cementation were nominated sub-contractors employed to design and carry out piling works. Because of poor ground conditions, difficulties were experienced in driving the 450 mm diameter piles. These difficulties, together with the admitted negligence of Cementation's operative, led to damage to an adjoining building. As a result, the contract was delayed while a revised piling scheme was devised, utilising 300 mm diameter piles. Cementation accepted that they were liable to indemnify the Co-operative Society for the damage to the adjoining building, but refused to accept liability for further claims submitted by the Society, which were:

- the additional cost of the revised piling scheme;
- loss and expense and fluctuations paid to the main contractor because of the extension of time awarded; and
- the Society's consequential loss for the delayed completion.

The main contract was JCT 63 and Cementation, as a nominated sub-contractor, had entered into a collateral warranty (RIBA Grey Form). In the judgement, an important limitation of the collateral warranty was highlighted, which was that the agreement did not warrant that the sub-contractor would execute the work with due care. The following statement was made by one of the appeal judges with regard to the collateral warranty:

> Under clause A(1) Cementation warranted to the Society as to the exercise of all reasonable skill and care (inter alia) 'the design of the Sub-Contract Works insofar as the Sub-Contract Works have been or will be designed by the Sub-Contractor' … I do not construe the word 'design' in the direct warranty agreement as embracing the means of carrying out the design; I am not persuaded that 'design' includes 'build' in this context.

The implications of the judge's comments were that, although the wording of the collateral warranty required the sub-contractor to exercise reasonable care in carrying out the design work, this same duty of reasonable care did not apply to the actual execution of the work.

The issues raised in the case were complex and for some reason the legal representative for the Society decided to base the appeal upon damages in tort and not upon breach of contract. Cementation were able to present a successful argument that the Society's costs were in fact 'economic loss' and therefore not recoverable under tort. However, during the course of this appeal, an important statement was made that, if there had been a contractual obligation on Cementation to carry out their subcontract works with due care, then it was accepted that Cementation, for breach of that contractual term, would be liable for the economic loss suffered by the Society. In recognition of this case, the JCT introduced a requirement into the SBC/Q to the effect that:

> The Contractor shall carry out and complete the Works in a proper and workmanlike manner... (clause 2.1)

To ensure that this requirement may be readily enforced, the SBC/Q clause 3.19 allows the A/CA to issue an instruction where there has been a failure to carry out the work in a proper and workmanlike manner, e.g. shuttering badly supported scaffolding in an unsafe condition, welding in dangerous circumstances causing a hazard to third parties, excessive piling vibration, etc.

Retention

One area that can have a significant impact on the employer's or contractor's cash flow is retention. Under clause 4.10.1, the employer is allowed to deduct a specified percentage from the monies due to the contractor following the issue of an interim certificate. To give the contractor some financial protection, the employer becomes a trustee in relation to the retention monies, with the implication that the monies are to be held in trust until such time as they are to be released to the contractor. But how secure is the retention money and how effective is its trust status? The trust status of the retention monies can be totally effective only if the retention money can be clearly identified and is kept separate from the employer's other accounts. For example, if an employer were to go into liquidation, the liquidator would refuse to hand over any retention money to the contractor unless the monies were clearly identified and kept in a separate account. These were obviously the issues concerning the contractor in the case of Rayack Construction Ltd v Lampeter Meat Co. Ltd (1979).

This case was based on a contract similar to the 1963 edition of the JCT contract with quantities. Rayack had tendered for the construction of a meat processing plant, with the retention percentage set at 50%, and a defects liability period of 5 years. An obvious reason for the imposition of such excessive terms would be to improve the employer's cash flow by severely reducing the amount of monies that would have to be paid to the contractor through interim certificates and by delaying the final release of retention for at least a further 5 years after practical completion.

The contractor became concerned at the amount of retention held by the employer and its lack of security; if the employer were to get into financial difficulties (and the contract requirements imposed would seem to imply a cash flow problem), the contractor might never recover his retention monies. During the progress of the project, therefore, the contractor sought a court order to have the monies, which at that time were in excess of £360,000, placed into a fund separate from the employer's normal funds.

The contractor was successful in obtaining a mandatory injunction requiring the employer to pay the retention money into a separate bank account and thereby secure its trust status. The following important statement was made:

> ...it is in my judgement clear that the purpose of the provisions for retention under the terms of condition 30(4)(a) is to protect both employer and contractor against

the risk of insolvency of the other. The employer is protected by his right to retain a proportion of the sum certified as due in respect of work done against the risk that claims in respect of any failure to carry out the architect's instructions or in respect of delay or other breaches of the contractor's obligations will, in the event of the contractor's insolvency, rank as unsecured debts. The contractor is protected by the provision of condition 30(40)(a) against the risk that his claim for payment of monies retained by the employer will similarly rank as an unsecured debt…. Thus, both are protected if and to the extent that the employer carried out his obligation to set aside as a separate trust fund a sum equal to the retention monies.

As a direct result of the Rayack case, the JCT introduced a provision allowing the contractor to request the employer to place the retention into a separate account (SBC/Q, clause 4.18.3). Obviously, from the employer's viewpoint such a request would have an impact upon his own cash flow and working capital. To overcome this problem, is it possible for an employer to delete the clause? This question arose in Wates Construction (London) Ltd v Franthom Property Ltd (1991).

In this case, the project had been let on JCT 80 with quantities but the contract had been amended by the deletion of the clause allowing the contractor to request that the retention be held in a separate account (i.e. clause 30.5.3). After the certificate of practical completion had been issued, the contractor requested that the remaining retention monies be placed in a separate and identifiable bank account. Franthom refused the request, on the basis that the operative clause had been deleted from the standard form of contract and the contractor's interests were being protected by the fact that the architect's certificates included a statement of retention that clearly identified the amounts of retention being withheld from the contractor.

In the Court of Appeal, it was held that the architect's statement of retention was not an adequate means of identifying and setting aside the retention monies. The deletion of clause 30.5.3 was not sufficient of itself to show clearly that it was the intention of both parties that the requirement for a separate trust did not exist. Clause 30.5.1 (i.e. 'the Employer's interest in the Retention is fiduciary as trustee for the Contractor…') stated clearly that the retention money was in fact to be held in trust by the employer. Wates were successful in their request for summary judgement for an injunction and Franthom were required to place the retention monies in a separate bank account, despite the deletion of clause 30.5.3.

This case demonstrates clearly the dangers of making amendments to standard forms in a piecemeal fashion. Any changes to the conditions need to be followed through to ensure that no ambiguity can arise through the interpretation of any associated clauses. When an employer does alter a standard form and an ambiguity arises as a result, then, under the 'contra proferentum' rule, the interpretation will normally be in favour of the other party.

In 1991, two more cases came to the courts involving different projects and different JCT forms of contract.

In the case of J F Finnegan Ltd v Ford Sellar Morris Developments Ltd (1991) (based on a JCT 81 Design and Build contract), the contractor (Finnegan) had been late in completing the works and the employer had deducted £50,165.61 in liquidated damages, having given all the necessary notices. Fifteen months after practical completion, Finnegan requested that the retention money be placed in a separate bank account. Upon the employer's refusal, Finnegan sought an interlocutory mandatory injunction (i.e. an interim injunction that may be granted pending the outcome of a case) to force them to comply. Ford Sellar's refusal was based on the following issues:

1. Finnegan should have made a request prior to the payment of each certificate; they could not make a global request at the end of the contract.
2. Finnegan had taken an unreasonable length of time to make their request. A mandatory injunction is an exceptional form of relief and it would not be appropriate in these circumstances.
3. The retention fund had ceased to exist because it had been used to reimburse the employer's liquidated damages.

The judgement was as follows:

1. The contractor may make a request for a separate trust fund at any time and he is not required to make repetitive requests.
2. As a consequence of the above, Finnegan were not guilty of delay; also, on balance of the circumstances there would be a greater risk of injustice through not granting a mandatory injunction, therefore an injunction would be granted.
3. Because this was a Design and Build contract, the notices relating to delay and liquidated damages were issued by the employer through the employer's representative, whose judgement could not be considered to be impartial. As a result, it was held that these notices did not have the same effect as those of an architect, which are binding on the parties until they are set aside by an arbitrator. Finnegan had a genuine dispute regarding the deduction of liquidated damages and the employer was required to remain trustee for the retention monies until the dispute was heard and was not entitled to appropriate them through the deduction of liquidated damages.

The second case of J F Finnegan Ltd v Ford Sellar Morris Developments Ltd (1991) related to a different project (let on JCT 80 with Approximate Quantities) and was heard separately from the one above but the arguments put forward were basically the same. There was no dispute relating to liquidated damages in this instance, although there was the added complication of the employer's assignment. Not surprisingly, the outcome mirrored the above decisions, and Finnegan successfully gained an injunction against the employer.

Retention and Employer's Insolvency

In the same year, another case was heard, which highlighted the problems that can arise when the employer faces insolvency. In MacJordan Construction Ltd v Brookmount Erostin Ltd (1991), the contract was let under JCT 81 with Contractor's Design. The contract was amended by deleting clause 30.4.2 (i.e. the requirement to set up a separate account if requested). Funding for the project was provided by a bank, who secured their loan by means of a fixed charge on the site and a floating charge on Brookmount's assets.

Brookmount were in financial difficulties when, on 4 March 1991, the bank appointed an administrative receiver and MacJordan, in accordance with the terms of the contract, subsequently determined their employment. The receivers issued an affidavit to the effect that Brookmount were insolvent. In an effort to salvage some funds from the project, MacJordan sought an interlocutory injunction to force Brookmount to set up a separate account for the retention money, even though the contract condition allowing this request had been deleted. On the basis of Wates v Franthom and Finnegan v Ford Sellar it would appear that this action should have been successful. Unfortunately, MacJordan failed to obtain the injunction. The court's decision was that, upon the appointment of the administrative receivers, the bank's rights, under its fixed and floating charge, took precedence over the contractor's right to request a separate trust fund. If the contractor had submitted the request prior to the appointment of the administrative receivers, his retention monies should have been secured.

Injunctions v The Balance of Risk

The contracting fraternity suffered another setback the following year during the hearing of GPT Realisations v Panatown (1992). The dispute referred to three projects that had been let on the JCT Design and Build form. The contractor (GPT) had received certificates of practical completion on two of the projects but on the third their employment had been determined. GPT had commenced arbitration proceedings on two of the projects and were claiming additional monies from the employer, who in turn was claiming monies from the contractor. An added complication was that GPT subsequently went into receivership and the receivers very belatedly sought an injunction requiring Panatown to place the retention in a separate account.

The receiver was not successful in his action and the judgement reinforced the aspect of fairness and risk that was touched upon in the case of J F Finnegan Ltd v Ford Sellar Morris Developments Ltd. The judge declared that he had to decide which course carried the lesser risk of injustice if, at a later stage, as a result of litigation, it became apparent that a wrong decision had been made at the time of granting the injunction. To some extent, this seemed to be contrary to the reasoning in Finnegan v Ford Sellar Morris, where an employer's right of access to retention monies was weakened because his claim for liquidated damages, under a Design and Build contract, was not supported by a disinterested party (i.e. the employer's representative acts on behalf of the employer and any decisions made by him are likely to be construed as biased). However, in this present case, it was considered that Panatown's claims against GPT were well founded:

> Here the defendants do have substantiated claims which overtop the amount of the retention money. The claims are substantiated by affidavit evidence which is both literally and metaphorically weighty.

In reaching the judgement the following factors of risk and fairness were considered:

1. The employer had a claim of set-off against the contractor.
2. The employer would lose the use of the money from his business.

3. The bank who appointed the receiver would not guarantee the contractor's liabilities if it was later proven that the injunction was wrongly granted.
4. The employer would be forced to pursue his claim against an insolvent company.

The judge's decision, therefore, was that the lesser risk lay with refusing to grant the injunction.

The key points determined by these cases are that, where a JCT contract bestows trust status on retention money:

- If the clause requiring the retention to be placed into a separate account has been deleted, the contractor's right to a separate account has not necessarily been extinguished.
- A contractor can request that retention monies be placed into a separate account at any time, even after the issue of the practical completion certificate (the courts will, however, consider the level of risk relating to the circumstances of the parties when asked to grant an injunction).
- Where an employer is the subject of insolvency proceedings, the contractor has lost his right to request a separate account for the retention monies; therefore, a contractor should consider requesting a separate account at the start of the project.
- Under the JCT Design and Build contract, where there is a genuine dispute regarding the deduction of monies from the contractor, the employer has no right of recourse to the retention monies.

Section 5: Payment

Variations

Where a quantity surveyor (QS) is required to value variations he must comply with the valuation rules contained within clauses 5.6 to 5.10. Under clause 5.6.1.1, if the variation work is 'of similar character to, is executed under similar conditions as, ... work set out in the Contract Bills' then the work will normally be valued in accordance with the rates provided in the contract bills. Under clause 5.6.1.2, the rule is that where work is 'of similar character to work set out in the Contract Bills but is not executed under similar conditions', the rates in the contract bills will provide the basis for the valuation but they are to be adjusted to take into account the effect the change in conditions would have on the work. It can be seen, therefore, that the conditions in which the variation work is carried out will have an impact upon how the work is to be valued, but what is meant by the term 'similar conditions'? An answer to this question may be found in Wates Construction Ltd v Bredero Fleet Ltd (1993), a dispute based on the 1980 edition of the SBC/Q. Some variations had arisen in the substructure works and the parties were in disagreement as to how the work should be valued. In the subsequent arbitration, the arbitrator was asked to determine the meaning of 'conditions' within the context of the valuation rules. The arbitrator's decision was that the term 'conditions' encompassed a broad range of issues, such as the knowledge the contractor would have presumably acquired throughout the pre-tender procedures and subsequent negotiations and the pricing of the cost plan, along with the express terms of the contract, physical conditions relating to time, the contract period, the nature of the works and any other matters that may have an impact upon the working environment. On appeal to the courts the arbitrator's definition was not upheld. The judge considered that the arbitrator had made an error in determining the meaning of 'conditions' and should not have taken into account any evidence beyond the contract documents. The judge stated that the term 'conditions' used in the valuation rules relate to the conditions that could be determined by reference to the contract bills, drawings and other contract documents. It can be seen, therefore, that where variation work has been executed it may be compared with the contract documents to determine whether or not the work has been carried out under similar conditions.

Under clauses 5.6.1.1 and 5.6.1.2, the rates in the contract bills provide the basis for valuing a variation. However, what happens if a contractor makes a mistake when submitting a bill rate? Should the 'incorrect' bill rate be used for valuing a variation or should the rate be corrected before valuing the work? This question arose in the case of Henry Boot Construction Ltd v Alstom Combined Cycles Ltd (2000). The project was let on an Institute of Civil Engineers (ICE) standard form of contract and the dispute related to the use of an erroneous rate in the valuation of a variation. Henry Boot had made a mistake when setting out the price for temporary sheet piling, which meant that the rate for the sheet piling was higher than was intended by the parties. When a variation was issued requiring additional temporary sheet piling, Henry Boot argued that the contract rate should be used to value the work, whereas Alstom claimed that the erroneous rate should be ignored and be replaced by a fair valuation. The arbitrator agreed with Alstom's argument, and decided that the work should be priced on the basis of a fair valuation. On appeal to the Technology and Construction Court, the judge's decision was that the contract rate was to be used to value the variation, regardless of the fact that the price had been submitted by mistake. Alstom appealed this decision but the Court of Appeal upheld the previous judgement. Although this case related to an ICE form of contract, the same principles would apply to the SBC/Q.

Clause 5.6.1.2 identifies that if an item of variation work is 'not of a similar character to work set out in the Contract Bills' then it should be valued using 'fair rates and prices'. But what is meant by 'fair rates and prices'? Does it mean that the contractor should be paid his reasonable costs for carrying out the works plus a fair allowance for profit and overheads? The case of Norwest Holst Construction Ltd v Co-operative Wholesale Society Ltd (1997) provides some advice on how such variation work should be valued. Norwest Holst was the contractor and the Co-operative Wholesale Society was the sub-contractor on a project to construct a library for John Moores University. A number of disputes had been referred to arbitration. Following the arbitrator's award both parties appealed to the court for judicial intervention concerning a number of issues. One of the questions raised by both parties concerned the method to be used for valuing some of the works. The answers provided by HHJ Thornton were: 'if the remeasured quantities cannot be valued in accordance with SR1, SR2 and SR3 (i.e. the contract rates), they are to be valued at fair rates and prices which mean "fair" in relation to the original sums shown in the schedule of rates SR1, SR2 and SR3.' He also went on to state that '. . . If there is a change in the scope of the sub-contract work which involves different work it may be necessary to create new rates for new items but these rates should be, where possible analogous to the existing rates in SR1, SR2 and SR3…'. What this means is that where fair rates and prices are to be calculated they should reflect the level of pricing used in creating the original contract rates. For example, if it could be shown that a contractor had priced a project at cost, with no allowance for profit and overheads, any variation work priced at fair rates and prices must reflect the same approach, i.e. the work must be priced at cost with no allowance for profit and overheads.

Fluctuations

Schedule 7, paragraphs A.5 and B.6 state that the QS and contractor may agree what is deemed to be the net amount payable under the fluctuations option A or option B. The wording of these paragraphs is slightly ambiguous and can lead to some confusion, as demonstrated by John Laing Construction Ltd v County and District Properties Ltd (1982). A dispute had been referred to arbitration and, as part of the procedure, the arbitrator raised a number of questions, which were to be referred to the courts. The works comprised a shopping development and was let on a JCT SBC/Q. The contract conditions incorporated a clause allowing for the recovery of 'full fluctuations' (i.e. as option B in the 2005 version of the SBC/Q). During the progress of the project the QS and contractor had agreed the value of fluctuations incurred, which were then included in the interim certificates and paid for by the employer. Unfortunately, it came to light that the contractor had failed to provide the architect with the appropriate written notices advising of events that would result in a claim for fluctuations (e.g. see fluctuation option paragraph B.5.1). The contract conditions stated that there would be no fluctuations payment made unless this notice was properly delivered. As a consequence the employer (County and District Properties) was looking to reclaim the fluctuation payments that had been made to the contractor. Laing argued that the contract conditions gave the QS the authority not only to agree the amount of fluctuations payable, but also to waive the requirement for the contractor to provide a fluctuations notice. The current wording used in the SBC/Q is as follows:

> The Quantity Surveyor and the Contractor may agree what shall be deemed for all the purposes of this Contract to be the net amount payable to or allowable by the Contractor in respect of the occurrence of any event such as is referred to in any of the provisions listed in paragraph B.5.1.

The judge held that the clause referred to by Laing gave the QS the authority only to agree the value of the net amount of fluctuations payable. The QS had no right to amend or ignore the contractual requirements relating to the fluctuation procedures, i.e. he could not waive the notification procedures, which were a condition precedent to a payment being made. Consequently, the employer was entitled to recover the fluctuation payments that had been improperly paid in the interim certificates.

Insurance

Any building project will bring together a number of parties, e.g. client, architect, engineer, contractor, sub-contractor, supplier, local authority and statutory undertaker, who will be involved in various stages of the works. The development of the project will involve the expenditure of thousands, if not millions, of pounds and will incur a high level of financial risk. Combined with this financial risk is the far greater risk to health and safety on site. It is an unfortunate fact that the construction industry tends to have a poor safety record compared with other industries. Because of the nature of the work undertaken there is a high risk of accidents causing injury or death to personnel, and damage and destruction to property. Recent legislation regarding health and safety is placing increasing pressure on members of the project team to become more aware of their responsibilities. Failure to comply with such legislation can lead to the imposition of considerable fines and perhaps to further civil action for compensation. In the case of personal injuries, the courts are now tending to award considerable sums of money against negligent parties. It is therefore vital that all parties to a construction project have the relevant insurances in place to cover these risks, and it is equally vital that the parties fully understand their liabilities and responsibilities in relation to the insurance procedures and provisions.

Unfortunately, the contract conditions relating to the project insurance requirements and the insurance documentation itself can appear very confusing and complex. As a result, it is not uncommon for parties to be unaware of their insurance liabilities, or for them to become embroiled in disputes as to the detail of the insurance provided and in disagreements as to who and what is covered by the various policies that have been put in place. Failure to understand or comply with policy requirements or contractual conditions may lead to insurance cover being removed or severely reduced. The consequence of such actions could lead to enormous financial loss and possibly insolvency. In the following section, a number of legal disputes are reviewed to highlight some of the insurance-related problems that have arisen in the past and the decisions that have been reached by the courts.

Employer's Liability, Clause 6.5

The employer as owner or occupier of land has a responsibility to avoid causing damage to adjoining or other properties. Because of the nature of building works, there is always the possibility of an event arising that may lead to such damage. The contractor will have provided the employer with an indemnity to cover damage to third party property, but the indemnity will only become operative where it can be shown that the damage resulted from the contractor's negligence, fault, etc. Consequently, there is the possibility that damage to adjoining properties may occur for which the employer may not be insured.

This fact was demonstrated in Gold v Patman & Fotheringham Ltd (1958). In this case, the contractor (Patman & Fotheringham) was involved in a piling operation, which damaged an adjoining property. There had been no negligence on the part of the contractor; the damage had arisen as a result of a weakening of support. The owner of the adjoining property successfully sued Gold (the owner of the property being renovated) on the basis that he had a right of support from Gold's land. The employer attempted to recover his costs from the contractor, who had been required to take out a public liability (i.e. third party) insurance. However, the policy covered the accidental negligence of the contractor only (the same provisions apply in the SBC/Q) and not that of the employer, thus the employer was uninsured for such an event.

As a consequence of that case, the JCT introduced a clause to cover such an event. The appropriate insurance is an optional requirement contained within clause 6.5.1. It is a joint names policy, to be taken out to cover expenses, loss, etc. that might be incurred by the employer as a result of injury or damage to any property. The events to be covered must arise from the execution of the works, and are listed as:

1. Collapse
2. Subsidence
3. Heave
4. Vibration
5. Weakening or removal of support
6. Lowering of ground water

On the face of it, such insurance would be covering a very wide range of risks, but the contract sets very strict limitations on the cover required. In fact, during the 1990s, the Insurance Market did not always provide policies that fully complied with the contract requirements current at that time. Because of extensive exclusions, the cover provided was less than that identified in the contract; consequently, the Association of British Insurers provided the JCT with a list of model exclusion clauses and, as a result, the insurance clause was redrafted, through the publication of JCT 80, Amendment 16, to mirror these model exclusion clauses. As a result of the amendment, the number of events that were excluded from the policy was increased from five to nine.

Works Insurance

Specified Perils

Although the contract provides a list of the risks that should be covered under the term 'specified perils', there can still arise a degree of confusion as to whether or not a particular event that has occurred on site is actually covered by the policy. The following two cases illustrate the difficulties encountered by parties in interpreting the term 'specified perils' where some of the works has suffered damage. The first, William Tomkinson and Sons Ltd v PCC of St. Michael in the Hamlet (1990), involved restoration works to a church. The project was let on the JCT Minor Works form of contract and the liability for the insurance of the works was at the sole risk of the employer. The contractor created a small inspection hole in the roof of the church and left it uncovered over the weekend. During this period, there was a heavy storm and a considerable amount of water penetrated the building through the inspection hole, causing damage to the furnishings and, in particular, to the church organ. The contractor contended that the responsibility for the remedial costs lay with the employer, as they carried the sole risk with regard to loss and damage by fire, lightning, explosion, storm and tempest. The decision of the court was that a heavy downfall of rain did not come within the definition of a storm for the purpose of perils insurance. The contractor was held responsible.

The second case, Computer and Systems Engineering plc v J Lelliot (Ilford) Ltd & Others (1990) involved a project to carry out work in the existing building of Computer and Systems Engineering. The works were let on the JCT 80 form of contract containing the pre-1986 insurance conditions. During the progress of the works, a sub-contractor (Stoddart Ltd) allowed a metal purlin to fall and damage the sprinkler system. The employer's property suffered considerable damage through the ensuing water escape and it was claimed by the contractor and sub-contractor that the employer was responsible for the cost of the damage. The contract provided, through clause 22C.1, that the employer was to carry the sole risk of damage to the existing building and property for perils. The perils definition included items for '… flood, burst or overflowing of water tanks, apparatus or pipes…'. In the judgements, it was held that the wording had the following implications:

1. Flooding was defined as being caused by 'rapid accumulation or sudden release of water from an external source, usually but not necessarily confined to the results of a natural phenomenon such as storm, tempest or downpour' (but see William Tomkinson & Sons Ltd);
2. The bursting or overflowing of a pipe referred to an internal bursting of pipes.

The court's interpretation of the contract wording meant that the water damage caused by the impact of the steel purlin was not covered by the employer's insurance. The damage could not be viewed as a flood and it was not caused by a bursting of a pipe and, as a result, the contractor was held responsible for the costs to the employer. In the current SBC/Q, there has been a change in the wording, to the effect that specified perils includes loss or damage caused or arising from 'escape of water from any tank, apparatus or pipes'; it is possible that under this new wording the damage would be covered by the employer's insurance. Interestingly, if this incident were to arise on a new build project, the contractor would be fully covered by the all risks insurance as this policy provides cover for impact damage.

Joint Names Policy

In a number of situations the contract conditions require that an insurance policy is to be a 'joint names policy'. The contract conditions provide a definition of a joint names policy, but this definition was amended in 1996. In the light of a court decision, the JCT decided that it was necessary to emphasise and clarify which parties were entitled to a benefit under the all risks insurance and who, as a result, were to be protected from an insurer's right of subrogation. An expanded definition of a joint names policy was therefore provided:

> a policy of insurance which includes the Employer and the Contractor as the composite insured (then follows the new wording) and under which the insurers have no right of recourse against any person named as an insured, or, pursuant to clause 22.3 recognised as an insured thereunder.

The reason for having a joint names policy is that, not only does the policy clearly cover and protect both employer and contractor, but it also prevents the insurance company from recovering any loss from one of the parties if they are responsible for the damage caused. Such a remedy is normally available to an insurance company through the 'Right of Subrogation', which allows the insurance company to sue, through the policy-holder, the party who was responsible for causing the loss, and thereby recover their costs and insurance monies paid out. For example, if the contractor had insured the works and the employer accidentally set fire to it, the insurers, through subrogation, could seek to recover their costs from the employer. However, if a joint names policy is in existence, the insurance company cannot recover costs from a policyholder.

The principle of subrogation was illustrated in the case of Petrofina (UK) Ltd & Others v Magnaload & Others (1984), which was a dispute relating to an oil refinery. During the process of the works, a sub-contractor (Magnaload) employed a subsub-contractor to hoist equipment into place. At some point, the hoisting equipment fell, causing substantial damage to the works (in the region of £1.25 million). The insurance company, through the employer (Petrofina), attempted to recover their costs by suing both Magnaload and the subsub-contractor. The insurance company had issued an all risks policy to Petrofina, as the employer, which also included the contractor and sub-contractor as insured. The insurance company were unsuccessful in their action against the sub-contractor (Magnaload) and the subsub-contractor. It was stated that, as a fundamental principle of law, an insurer cannot sue a co-insured by way of subrogation.

Recognition of Sub-contractors

With reference to the all risks policies, it is a contract requirement (clauses 6.9.1.1 and 6.9.1.2) that the policy must recognise each sub-contractor as joint insured or, alternatively, the policy may be endorsed to signify that the insurers will waive any subrogation rights they may have against the sub-contractors. It is important to note that a sub-contractor's benefits under this policy are limited to 'Specified Perils' only; they do not have the full benefit of the all risks policy. Apparently the insurance companies were not prepared to extend the full all risks cover to sub-contractors; they were concerned at the level of risk involved with certain sub-contractors with regard to the theft of valuable goods and equipment.[6] It can be seen, therefore, that sub-contractors will enjoy the benefit of the specified perils cover and do not need to be concerned about the insurance companies' right of subrogation, but do subsub-contractors enjoy the same benefits? When the JCT published the 1998 edition of the SBC/Q and the associated subcontract conditions (DSC/C) they were of the opinion that subsub-contractors were not covered by the specified perils policy, as evidenced by the following statement:

> It should be noted that, since the benefit of the Main Contract Joint Names Policy is only extended to 'Domestic Sub-Contractors' as defined in clause 19.2 of the Main Contract Conditions, such benefit only extends to the Sub-Contractor himself and not to any company, firm or person to whom the Sub-Contractor sub-lets any portion of the Sub-Contract Works.[b]

This opinion was certainly supported by the case of City of Manchester v Fram Gerrard Ltd (1974). The case involved an early version of the JCT standard form of contract and

---

[b]Footnotes [p], [p$_1$] and [p$_2$] to DSC/C sub-contract conditions.

part of the dispute revolved around whether the term 'sub-contractor' should also include subsub-contractors. It was held that reference to sub-contractors in the contract conditions did not extend to cover subsub-contractors. However, in the more recent case of Petrofina (UK) Ltd and Others v Magnaload Ltd and Others (1983) a completely different opinion was expressed. This project was based upon an engineering contract, not a JCT form, and the dispute concerned subrogation rights under a contractor's all risk policy and whether a subsub-contractor had the benefit of the all risks insurance. As part of the judgement it was held that 'as regards Magnaload…although strictly speaking they were sub-sub-contractors the word "sub-contractors" must include sub-sub-contractors as well as sub-contractors.' Both of these cases were heard in the lower courts and, until a similar case is taken to the higher courts, this issue will remain a 'grey area'. In the circumstances, it would be advisable to consider agreeing the definition of 'sub-contractor' with an insurer at the outset. It is interesting to note that the JCT has removed its comment concerning subsub-contractors from the current subcontract conditions.

Damage to Property

Through clause 6.2, the contractor indemnifies the employer for certain claims that may arise as a result of damage to property as a result of the works being carried out. However, the contractor's liability for injury or damage to property does not include the works and site materials and, where the project relates to an existing building, does not include the existing structures and contents that are covered by the employer's insurance under paragraph C.1 of insurance option C. The opinion is that, as these items are already covered by works insurance found elsewhere in the contract conditions, it would not make sense to duplicate the cover. The wording originally used by the JCT to achieve this outcome was:

> The Contractor shall, subject to clause 20.3 (now 6.3) and, where applicable, clause 22C.1 (now paragraph C.1), be liable for, and shall indemnify the Employer against, any expense, liability, loss, claim or proceedings in respect of any injury or damage whatsoever to any property real or personal in so far as such injury or damage arises out of or in the course or by the reason of the carrying out of the Works…

Unfortunately, because of the wording used above, it was not always clear whether a contractor's indemnity had been effectively limited as the JCT intended, especially on projects involving work to existing buildings.

A number of court cases[c] have addressed the issue of whether a contractor or subcontractor may have a duty of care towards the employer with regard to damage caused to the works or to the employer's existing structure. Argument and debate have ensued regarding the implications of the employer having insured against these damages, questioning whether exemption terms incorporated in the main contract may be relied upon by a sub-contractor and whether there was a recognition or waiver in

---

[c]Scruttons Ltd v Midland Silicone Ltd (1962); Norwich CC v Harvey (1988); Ossory Road v Balfour Beatty (1993); British Telecommunications plc v James Thomson & Sons Ltd (1998).

the insurance policy for the benefit of sub-contractors. The outcomes of the cases have frequently been at conflict with one another, usually on the basis that 'Each case must turn…on its own facts,…'[d]

The uncertainties associated with the contract wording were illustrated in the case of Ossory Road (Skelmersdale) Ltd v Balfour Beatty Building Ltd and Others (1993). The project was for the construction of a retail outlet and car park, and the refurbishment of public areas. The contract used was JCT 80, including Amendment 6, and the works insurance was to be taken out by the employer (Ossory) through clause 22C. A fire occurred, which caused extensive damage to the new works but did not directly affect the existing structure. Ossory claimed that the fire was the result of negligence by the domestic roofing sub-contractor (Briggs Amasco) for whom Balfour Beatty was responsible. Ossory sought damages from both Balfour Beatty and the domestic sub-contractor.

With regard to the damage to the new works, the judge held that (in line with previous cases) Balfour Beatty was not liable to Ossory for damage caused by a specified peril, even though it may have been caused by Balfour Beatty's negligence. Balfour Beatty was not liable because Ossory was required to take out a joint names all risks insurance for the new works, and this express term removed any liability Balfour Beatty may have had to Ossory. It was held that the employer had accepted the risk of fire damage, etc. through the existence of clause 22C, which required him to insure against such an event.

Although the fire damage suffered by Ossory was limited to areas of the new works, the judge in his summing up also endeavoured to identify what liability Balfour Beatty (the main contractor) may have had if damage had occurred to the existing structure and contents. He commented that in clause 20.3 (see clause 6.3.1) the draftsman had referred only to excluding the contractor's liability for 'the Works, works executed' whereas if wording such as 'the Works or any part thereof' had been used, then the existing structure would have been clearly excluded from the contractor's indemnity. Consequently, the judge had to consider the implications of clause 20.2 (see clause 6.2), where it makes reference to the contractor's liability being subject to clause 22C.1 (now Schedule 3, paragraph C.1):

> …the difficult question is whether the words 'subject to…clause 22C.1' mean that Ossory bore the whole risk of damage to the existing structure caused by a Specified Peril whether or not the damage was caused by the negligence of Balfour.

The problem facing the judge was that clause 22C.1 was an insurance clause and not an indemnity clause and he therefore had to consider whether reference to clause 22C.1 implied that a contractor's indemnity is limited by the insurance that the employer is required to take out. The judge commented that,

> It would have been a simple matter to have stated in clause 20.3 that property real or personal did not include an existing structure when the damage or injury to it was caused by a Specified Peril.

---

[d]Garland J in Norwich CC v Harvey (1989).

Eventually the judge held that the reference to clause 22C.1 in clause 20.2 would in fact limit Balfour Beatty's indemnity to Ossory. However, his judgement did not give wholehearted support to the wording used by the JCT: 'After some hesitation I have reached the conclusion that in respect of a fire to an existing structure caused by Balfour's negligence in the carrying out of the works Balfour would be exempted for all liability to Ossory and their obligations to indemnify Ossory against third party claims.'

In response to the criticisms levelled at the wording of clause 20.2, the JCT issued an amendment in July 1996 that attempted to remove the uncertainty and ambiguity associated with the then current wording. The new wording (which is along the lines suggested by Judge Fox-Andrews in Ossory v Balfour Beatty) more clearly defined the limit of the contractor's liability with regard to property when working on an existing structure or a refurbishment project. The current SBC/Q has slightly changed the wording from the 1996 amendment, but the meaning is still the same:

> This liability and indemnity is subject to clause 6.3 and, where Insurance Option C (Schedule 3, paragraph C.1) applies, excludes loss or damage to any property required to be insured thereunder caused by a Specified Peril.

As a result of the revised wording, it should now be evident that a contractor is not liable for specified perils damage to existing structures or property covered by the employer's insurance (i.e. Schedule 3, paragraph C.1), but what is the situation for sub-contractors who do not have the benefit of the employer's specified perils insurance for the existing works?

Sub-contractor's Liability?

In the case of Ossory v Balfour Beatty, it had been held that the contractor was not liable to the employer for fire damage to the new works, even though the fire had been caused by the contractor's (or his sub-contractor's) negligence. However, the judge had also been required to consider whether the domestic sub-contractor (Briggs Amasco) could be liable to the employer through a duty of care. Here, the judge followed the decision of an earlier case heard in the Court of Appeal (Norwich City Council v Harvey and Briggs Amasco, 1988) and held that the sub-contractor did not owe a duty of care at common law to the employer. 'I have held that the injury or damage to the Works or work executed was at the sole risk of Ossory as between Ossory and Balfour, under the provisions of clause 20.2. I have reached the conclusion that the same position pertains between Ossory and Briggs.'

To fully understand this judgement, it is worthwhile to briefly review the case of Norwich v Harvey and Briggs Amasco, which was a dispute similar to that between Ossory and Balfour Beatty. It was claimed that the roofing sub-contractor had negligently caused a fire, which resulted in damage to the works and existing structure. Under the main contract (JCT 63), the employer was responsible for insuring the existing structure and contents. The original trial judge acknowledged that there was no privity of contract between the employer and sub-contractor, but raised the following issue:

...is the duty owed by the defendant to the plaintiff qualified by the plaintiffs' contract with the main contractor, or to put it more broadly, by the plaintiffs' propounding a scheme whereby they accepted the risk of damage by fire and other perils to their own property – existing structures and contents...? I am left in no doubt that the duty in tort owed by the sub-contractor to the employer is so qualified.

The appeal court supported the approach adopted by the trial judge and concluded that, as a result of the contract wording, the employer had accepted the sole risk of damage by fire and that this was to include fire caused through a sub-contractor's negligence. It would appear, therefore, that a sub-contractor may not be liable for causing specified perils damage to the employer's property, even though the sub-contractor is not covered by the employer's specified perils insurance. This principle was tested a number of years later in the case of British Telecommunications plc v James Thomson & Sons Ltd (1998).

This dispute, concerning fire damage to existing structures, progressed from the lower court through to the Court of Appeal and eventually to the House of Lords, which illustrates the legal uncertainties surrounding some of the JCT insurance provisions. British Telecommunications (BT) employed MDW Ltd, under JCT 1980 (1988 edition), to carry out repair and refurbishment work to a switching station. The steelwork was sub-let to James Thomson & Sons, a domestic sub-contractor. During the progress of the project a fire broke out, causing damage to the switching station. BT held the view that the fire had been caused by the negligence of Thomson and sued for damages. BT failed in both the first instance and the subsequent appeal. Thomson's argument was that it would not be fair or reasonable to impose a duty of care upon a sub-contractor for injury to an employer when the employer was under a contractual obligation to insure for the above risks. The lower courts supported this approach and were not prepared to impose a duty of care on Thompson with regard to the damage suffered by BT. In fact, the following statement was made by one of the appeal judges:

If the sub-contractor is aware that the employer has undertaken to insure against the risk of negligence on the part of the sub-contractor, then the latter is entitled to assume not merely that he need not himself insure but that he is not under any duty of care to the employer with regard to any loss or damage caused by his actions.

BT pursued their claim to the House of Lords, where the decision of the lower courts was overruled. It was held that Thomson did owe a duty of care to BT and that the insurance requirements in the contract conditions actually reinforced this view rather than negating it, as expressed in the opinion above. The reasoning behind this decision was that, although BT were obliged to take out insurance under 22C.1 (now Schedule 3, paragraph C.1) to provide specified perils cover for existing property and structures, domestic sub-contractors were specifically excluded from having any benefit or recognition under the policy.

In the case of Norwich v Harvey, which was based on JCT 1963, it was held that the contractual wording, requiring the employer to insure (at the sole risk of the employer) the existing structure and contents from specified perils, in effect qualified a sub-contractor's potential duty of care to the employer. The employer had accepted the sole risk of fire damage. In Ossory v Balfour Beatty (based on JCT 1988), although the contract wording had by now been changed to remove the reference to the employer accepting the 'sole risk', it was held that the employer still had to accept this risk, thereby limiting a contractor's or sub-contractor's duty of care. It was also expressed that a sub-contractor's duty of care would not be affected by whether or not the sub-contractor received recognition under the insurance policy. However, in BT v Thomas, the House of Lords considered that the contractual wording in the 1988 edition of the contract (and in particular a domestic sub-contractor's lack of recognition in the policy) actually reinforced a sub-contractor's duty of care. This demonstrates the care that is needed when drafting contract conditions, as well as the need to draft the terms in a clear and unambiguous manner. This sentiment was reinforced by the editors of the Construction Industry Law Letter in response to the case of Ossory v Balfour Beatty: 'We intend no discourtesy to either the draftsman of the various contracts or to the insurance market when we say that it is a pity that these complex provisions are not much simpler and clearer.'[10]

Termination

Section 8 of the SBC/Q sets out the grounds and procedures for the termination of the contractor's employment. There are two important points to consider when operating these procedures: a notice of termination should not be given unreasonably or vexatiously, and the timing of the notices must be correct. Some advice regarding these points may be found by reference to JM Hill & Sons Ltd v London Borough of Camden (1980). This case is based on the 1963 edition of the JCT contract and the court had to consider whether the contractor had correctly terminated his employment. One of the questions considered was whether the contractor had acted unreasonably in issuing a termination notice.

Ormrod LJ provided his definition of the term 'unreasonable': 'I find it difficult, in the circumstances of this case, to imagine how the plaintiffs could be said to have behaved unreasonably in this regard. It seems to me that they had the most cogent reasons for taking advantage of such remedies as the contract gave them. But what the word "unreasonably" means in this context, one does not know. I imagine that it is meant to protect an employer who is a day out of time in payment, or whose cheque is in the post, or perhaps because the bank has closed…something accidental or purely incidental so that the court could see the contractor was taking advantage of the other side in circumstances in which, from a business point of view, it would be totally unfair and almost smacking of sharp practice.'

Another question that arose in this case was at what moment did the contractor's notice of termination actually take effect? In that edition of the contract (and the 1980 edition), reference was made to a notice of determination being served on the employer. Did it mean that the notice became effective on the day it was posted, or on the day it was received by the employer? The court's decision was that a notice would not become effective until it was received by the other party. The wording in the current SBC/Q has been slightly modified from the earlier editions so that it now states clearly that the termination becomes effective upon receipt of the notice.

Within Section 8 of the SBC/Q, the JCT sets out the grounds that allow the contractor's employment to be terminated. One such reason for terminating the contractor's employment is where the contractor 'fails to proceed regularly and diligently with the Works or the design of the Contractor's Design Portion'. In the past, there had been a degree of uncertainty as to what this phrase actually meant, and contract administrators had been wary of terminating a contractor's employment under this heading. A problem for contract administrators is that, under clause 8.2.1, a notice of termination must not be given unreasonably or vexatiously, i.e. where there are insufficient grounds to issue a termination notice or where the notice is issued out of malice. This problem was highlighted in the case of West Faulkner Associates v London Borough of Newham (1992).

The contract in question was let on JCT 63 and incorporated the standard amendments imposed by Newham Borough Council. The works involved the refurbishment of a number of properties being managed by a housing association. The contractor appointed to carry out the works (Moss) failed to maintain progress in accordance with his agreed programme. The first three blocks to be refurbished should each have been completed within 9 to 10 weeks. In fact, each block took the contractor nearly 9 months to complete. These delays caused considerable problems for the tenants, who had been decanted on the basis that they would soon be returned to their refurbished properties. The council also incurred considerable additional expense through the weekly compensation they were paying the tenants for having to live in unsatisfactory temporary accommodation.

During discussions between the architect (West Faulkner Associates) and the council, it appeared that the architect was of the opinion that the poor progress on site by the contractor did not constitute a breach of contract, i.e. a failure to proceed regularly and diligently with the works. The architect found himself in a quandary, as there had always been a difference of opinion as to the strict legal interpretation of the phrase 'regularly and diligently'; he had obtained outside advice, which tended to support the view that the contractor may not be in breach of contract. The reasoning behind this advice was that the contractor had sufficient resources on site to carry out the works and there was continuous activity; therefore, despite the contractor's glaring failure to maintain progress, the architect was unwilling to operate the termination procedure and would not issue a notice of default.

Eventually, the council took matters into their own hands and reached a settlement with the contractor. Moss was to be paid for the work in hand, but was not to start any fresh work. Once the contractor had completed the agreed works, the council terminated the architect's employment. The architect sued for his fees and the council counterclaimed for the additional costs they had incurred, which they attributed to breaches of contract committed by the architect. The council claimed the architect had committed a number of breaches, but the relevant one in this instance was that the architect:

…failed to use or give proper consideration to the use of the provisions of the contract enabling the Defendant to determine the Contractor's employment.

As already mentioned, one of the key points at issue was the interpretation of the phrase 'regularly and diligently'. In this case, the following definition was provided by Judge Newey:

…'Regularly and diligently' should be construed together…in essence they mean simply that the contractors must go about their work in such a way as to achieve their contractual obligations. This requires them to plan their work, to lead and to manage their workforce, to provide sufficient and proper materials and to employ competent tradesmen, so that the works are fully carried out to an acceptable standard and that all time, sequence and other provisions of the contract are fulfilled.

On the factual evidence I think it certain that Moss did not plan their work properly, did not provide efficient leadership or management and that some at least of their tradespeople were not reasonably competent in terms of speed and/or quality of work. They had adequate quantities of materials and…men on site but, without being efficiently applied that did not avail them.

As a consequence of the court's interpretation, it was held that the architect was in breach of contract for failing to determine the employment of Moss. In the judge's opinion, the contractor's failure to proceed regularly and diligently with the works was glaringly obvious, and the architect could have issued a notice of default. The court's decision was upheld in a subsequent appeal. As a result of this case, future contract administrators have the benefit of the court's decision and their interpretation of the meaning of 'regularly and diligently'.

Clause 8.8 of SBC/Q sets out the procedures to be followed where the employer decides he does not wish to carry on and complete the works after the termination of the contractor's employment. In earlier versions of the SBC/Q, there had been no similar clause. It must have been assumed that the employer would always wish to complete the project and consequently there were no contract conditions to cover the employer's abandonment of the works. The problems that could arise as a result of this non-completion were illustrated in Tern Construction v RBS Garages (1992).

The employer (RBS Garages) terminated the contractor's employment after the appointment of an administrative receiver to the contractor. At that time the works were only partially completed. The contractor had received payment under the first certificate but the employer had not honoured the second certificate, issued 6 days before the termination. The employer was correct in the action that he had taken as, under JCT 80, clause 27.4.4 (pre-Amendment 11); he was not obliged to make any further payment until a reasonable time after the completion of the works.

In this instance, the employer decided not to proceed with the project and claimed, for a number of reasons, that he was not obliged to make a payment to the contractor. Tern Construction claimed that they were entitled to receive £158,640 for work carried out under the two certificates, a further £27,926.06 for works carried out subsequent to the last certificate, and £27,990.31 for VAT.

Because there were no express contract conditions covering the event where the employer abandons the works, the payment provisions had to be implied by the courts. They had to decide what payments, if any, the contractor was entitled to receive and what damages, if any, the employer was able to claim. The outcome of the case was that the employer had to honour the interim certificates, which had been issued. The value of the certificates could be reviewed by the court in the light of any errors, e.g. overvaluation, defective work, etc., but the employer could not deduct from the certificates for any losses he had incurred as a result of the non-completion of the works. The employer was entitled to recover his losses, but not under clause 27. The following comment was made:

> In my opinion terms must be implied to the effect that when the Employer's expressly stated or by their conduct showed that they did not intend to complete the Works or when after a reasonable time they failed to re-start them, the bar on their liability to make payments to the Contractors must have lifted and days could pass for the purposes of clause 30.1. Equally I think that the Employer must have become entitled to recover any sums due to them from the Contractors in a manner other than that laid down in 27.4.4.

Fortunately for future employers, this problem has now been resolved through the inclusion of express contract conditions, which lay down the procedures to be followed and affirm the right of the employer to deduct his direct loss and/or damage caused by the termination.

Mediation

In recent years, the construction industry has given some consideration to the idea of using mediation as a means of resolving construction disputes before the parties become too entrenched and polarised in their arguments. This procedure has received the support of the JCT, who introduced a mediation clause into the 2005 edition of the SBC/Q. Unfortunately, the clause is very limited in content and procedure and at best is just a suggestion to the parties that they may wish to consider the use of mediation in settling any disputes that may arise. However, there is good reason why the JCT did not make mediation a mandatory procedure, as illustrated in the following cases. In R G Carter Limited v Edmund Nuttal Limited (2000), a standard form of subcontract (DOM/1) had been amended so as to require the parties to refer any differences to a mediator. The contract stated that, if no settlement was reached within 6 weeks of the mediator's appointment, only then may the difference be referred to adjudication. It was held that such a clause was a clear contravention of the HGCRA as it removed the sub-contractor's statutory right to adjudication and, as a result, the mediation clause was held to be invalid.

The above case related to a situation where a contractor had amended a standard form of subcontract. Would the court's decision be the same where similar terms were part of an unamended standard form? In John Mowlem & Company plc v Hydra-Tighe Limited (2000) the subcontract works was let on the Engineering Construction Contract Sub-contract Conditions (incorporating Option Y). Within the contract conditions relating to adjudication, it was stated that a dispute would not exist unless a 'Notice of Dissatisfaction' had been given and the issue(s) raised had not been satisfactorily resolved within 4 weeks of that notice. The effect of this clause is to delay a party's right to refer a dispute to adjudication. It was held that the clause did not comply with the requirements of the HGCRA and, as a result, the whole adjudication scheme within the standard form of contract was void and replaced with the adjudication procedures from the Scheme for Construction Contracts.

The situation at present, therefore, is that a form of contract may recommend the use of mediation and provide the necessary procedural detail if wished, but it may not impose the procedure on the parties.

References

1. Joint Contracts Tribunal. Amendment 15, RIBA Publications, July 1995, p. 4.
2. Oxford University Press v John Stedman Design Group, Construction Industry Law Letter, Legal Studies & Services Ltd, p. 591.
3. Parris, J. (1985). *The Standard Form of Building Contract JCT 80* (2nd ed.) (p. 180). Collins Professional and Technical Books.
4. Keating, D. (1978). *Building Contracts* (4th ed.) (p. 346). Sweet & Maxwell.
5. ·Madge, P. *A Concise Guide to the JCT 1986 Insurance Clauses* (p. 60). RIBA Publications.
6. Insurance Provisions, Construction Industry Law Letter, Legal Studies & Services Ltd, p. 884.

Legal References

All ER   All England Law Reports

BLR   Building Law Reports

CILL   Construction Industry Law Letter

CLD   Construction Law Digest

WLR   Weekly Law Reports

Index